Major Cities of Middle America

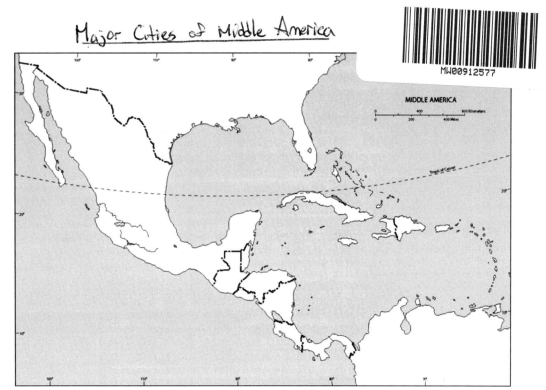

MIDDLE AMERICA

Study Guide

TO ACCOMPANY

Geography

Realms, Regions, and Concepts

Fourteenth Edition

STUDY GUIDE

TO ACCOMPANY

GEOGRAPHY

Realms, Regions, and Concepts
Fourteenth Edition

H. J. de Blij
John A. Hannah Professor of Geography
Michigan State University

Peter O. Muller
Professor, Department of Geography and Regional Studies
University of Miami

Prepared by
Justin Scheidt
Ferris State University

John Wiley & Sons, Inc.

Cover Photo: H. J. de Blij

To order books or for customer service, please call 1-800-CALL-WILEY (225-5945).

ISBN-13   978-0470-59828-3

Printed in the United States of America

10 9 8 7 6 5 4 3 2

Printed and bound by Bind-Rite Robbinsville.

INTRODUCTORY CHAPTER: WORLD TOPICS

SECTION I.1:   Map Creation

In this section you will create your own detailed maps of the different major physical, cultural, geopolitical and environmental attributes of the. Each of these maps is critically important for building your foundation of the spatial distribution of physical, cultural and economic world concepts. By creating these on your own using the blank maps provided at the end of this chapter, you will actively learn where these different features are located and how their spatial distribution affects the ways of life in each of these respective areas.

Blank Maps for The World are provided for you in Section I.3. Label each of your maps with the appropriate name and title. For instance, the first map you will create will be called 'Map I – 1: Continents and Oceans of the World' (The letter 'I' standing for 'Introductory Chapter').

Map I–1:   Continents and Oceans of the World

There are seven continents and four major oceans that comprise the surface of Planet Earth. To assist with creation of this map, refer to world map on page 1 of the textbook.

Continents of the World:
- North America   - South America   - Africa   - Asia
- Australia   - Europe   - Antarctica

Oceans of the World:
- Atlantic Ocean   - Pacific Ocean   - Indian Ocean   - Arctic Ocean

Map I–2:   Average Annual Worldwide Precipitation

Using Figure G – 6 (p. 14) of the textbook, you will create a map that shows the spatial distribution of world precipitation levels. The distribution (or lack thereof) has a substantial impact on not only the physical characteristics of a realm, but also upon its settlement and cultural patterns. These worldwide precipitation patterns are divided into seven categories below, based on inches of precipitation per year. Use a different colored pen/pencil for each category for clarity in your map.

Highest *Precipitation Levels in the World*:   Areas exceeding 200 inches/year

Very High *Levels of Precipitation*:   Regions averaging between 80–200 inches/year

High *Levels of Precipitation*:   Areas averaging 40–80 inches/year

Moderate *Levels of Precipitation*:   Regions averaging between 20–40 inches/year

Low *Levels of Precipitation*:   Areas averaging between 12–20 inches/year

Very Low *Levels of Precipitation*:   Regions averaging between 4–12 inches/year

World's Driest *Precipitation Regions*:   Areas averaging below 4 inches/year of precipitation

Map I–3:   Major Climate Regions of the World

The variation of climates around the world has contributed substantially to population distribution, agricultural patterns, and many more variables of our way of life. Using Figure G – 7 (pp. 16–17) as a guide, you will create a basic climate map of the world that shows where you will find the primary climate types in the Köppen System. In particular, you will identify only the general locations of the A, B, C, D, E and H climates. Refer to the textbook for an explanation of the symbols used in this climate classification system. Using a different colored pen/pencil for each category, outline and then color in these climate types. Draw a general circle around where you will find each of these climate types.

Tropical (A) Climates: A – Climates are along the Equator and are regions of hot weather, usually without a pronounced winter season.

Dry (B) Climates: B - Climates are typically located near 30°N or S of the equator, and represented by places such as the Sahara Desert and Australian Outback.

Mild Mid-Latitude (C) Climates: C – Climates are characterized by mild winters and mild to hot summers. Rarely will a C – Climate location receive snowfall in the winter season, especially consistent snowfall.

Cold Mid-Latitude (D) Climates: D – Climates are only found in the Northern Hemisphere, and are characterized by having four pronounced seasons throughout the year.

Polar (E) Climates: E – Climates are cold throughout the year, either existing as tundra or permanent ice regions, and do not have a pronounced summer season. These are found near the North and South Pole.

High (H) Climates: H – Climates account for the idea that tall mountain chains that have snow on them year-round.

Map I–4:   Extent of the Last Glacial Age on Earth

As recently as 16,000 years ago, massive glaciers covered much of the surface of the Northern Hemisphere of Earth. These glaciers left substantial scars across the landscapes of North America, Europe and Russia, resulting in a wide variety of newly-formed physical features and water supplies. For example, The Great Lakes of North America are one of the most prominent reminders of these glaciers.

Using Figure G – 5 (p. 12), you will create a map that shows both the previous and current extent of glaciers in the Northern Hemisphere. Use a different colored pen for each glacial extent. The spatial extent of world glaciation will be significant when comparing historical to present-day.

Historical Glaciation Extent: Draw a blue line on your map that shows how far the glaciers used to extend into North America, Europe and Russia. Be sure to also account for the isolated glaciers found in the Alps and Caucasus Mountains (Europe), and Present-Day California in the USA.

Present – Day Glaciation: Draw a red line on your map that encloses the present extent of the glaciers in the world today. Though not shown on Figure G – 5, also include the Arctic Ocean above 80 degrees North Latitude.

Map I–5:   World Population Densities

The human populations across Earth are unevenly distributed. Using the World Population Distribution map on Figure G – 8 (pp. 18–19), we will find that six general population density categories exist across the planet. However, for the sake of space and to obtain the general introductory ideas, we will only highlight those regions with either very high or very low population densities for now. Subsequent examinations of the world realms will give a more in-depth approach to population concentrations.

Using a different colored pen/pencil for each category, outline and color the countries/sections of countries that fit into each of these designations. Once completed, you will see a spatial pattern of population distribution emerge across Earth.

Extremely Dense: The most dense population category in the world. Presently these are regionally only found in Northeast and Southwest India; however, the Eastern coastline of China may soon qualify for this density status.

Very Dense: The very dense categories account for most of the clustered urban regions of the world, particularly:
- Boston to Washington D.C., on the eastern shores of the United States
- Eastern China, all of Japan and Taiwan, and the rest of India along with Pakistan.
- Eastern South America
- West – Central Africa
- Western Europe

Very Sparse: These regions will have almost no cities and very few inhabitants scattered across a vast expanse of land. Reasons for the lack of people could include inhospitable climate or terrain. Areas of the world that are the best examples of very sparse populations include the following:
- Northern Canada, more than 100 miles north of the USA border
- Greenland
- Northern Russia
- Western China and Mongolia
- The Australian Outback

Empty: These locations in the world will not have a permanent human population. They include the following:
- The North Pole regions
- Antarctica
- Southeast Saudi Arabia, an area called the Rub Al – Khali (more on this region in Chapter 7).

Map I–6:   *Worldwide Language Families*

Using Figure G – 10 (pp. 22–23), you will create a map that shows the spatial distribution of language families spoken across the world. For the purposes of this exercise, you will show the major language families across the world. Use a different colored pen/pencil (if possible) to detail where each of these are located. The brief descriptions below are meant as a general guide for locating each, and not as a substitute for referring to the map on pp. 22–23.

Indo – European Family:	Found in North America, South America, Europe, South Asia and Australia; these include English and the Romance Languages.
Afro – Asiatic Family:	Found across North Africa and the Middle East.
Niger – Congo Family:	Found in Central and Southern Africa primarily.
Sudanic Family:	Found in East-Central Africa, centered on the country of Sudan.
Khoisan Family:	Located in SW Africa; one of the oldest language families in the world.
Uralic Family:	Found in Northern Europe and Northwest Russia.
Altaic Family:	Scattered throughout the Middle East, Central Asia and Eastern Russia.
Sino – Tibetan Family:	Found in East Asia, contains the highest number of speakers in the world and includes variations of the Chinese (Mandarin) Language.
Japanese and Korean Families:	Located in the countries of Japan and Korea; very different languages than those found in East Asia.
Dravidian Family:	Located in Southern India and Eastern Pakistan.
Austronesian:	Found in the islands of Southeast Asia and the Pacific Ocean Realm.
Amerindian:	Located throughout North and South America.

Map I–7:   *States and Economies of the World*

Using Figure G – 11 (pp. 24–25) as a guide, you will create a map showing the spatial distribution of world economies and the income levels of these respective countries. Four categories are created to show this variability. Use a different colored pen/pencil for each group, and refer to the textbook for additional information on these locations. Examples of a country in each category will be given below; however, this is not an all-inclusive list. You will need to refer to pp. 24–25 to complete the map accurately.

High – Income Economies:	Examples: North America, Western Europe, Australia
Upper – Middle Income Economies:	Examples: Brazil, Mexico, Argentina, Russia
Lower – Middle Income Economies:	Examples: China, Thailand, Indonesia
Low Income Economies:	Examples: India, Central Africa

Map I–8: *Major World Environmental Issues*

Throughout this course and study guide, we will examine environmental issues that are either natural or human-induced, and affect the way of life for a substantial percentage of Earth's population. In this map, you will circle the areas below that have the most blatant and glaring environmental issues found on Planet Earth. These will be explored more as the semester progresses, both in the textbook and this study guide, and by no means is this list exhaustive- it is meant as a very small sample of the many environmental issues facing citizens of the world today.

These environmental issues are divided into categories below. Using a different colored pen/pencil for each, circle or highlight the areas and label them using their respective environmental situation.

Deforestation:	Worst problems in the world are centered in Northern South America, in the Amazon Rainforest. However, additional problems exist in Central Africa and Southeast Asia, particularly Indonesia.
Overgrazing of Grassland:	Across the world, lack of space is leading to herds overgrazing a grassland and not allowing it to replenish, changing that grassland to desert over time. This problem is the worst in North – Central Africa, in an area just below the Sahara Desert called the Sahel. It is also a severe problem in Argentina (South America).
Groundwater Depletion:	The Great Plains of the USA has seen tremendous problems with taking more water out of the ground than can be replenished over time. The primary reason is for irrigation of crops on land that cannot otherwise support agricultural activity. Groundwater depletion is also a very big problem across the country of Australia.
The Disappearance of Lakes:	Two examples illustrate this worldwide problem: • The Aral Sea (Kazakhstan, near southern Russia) has seen over 80% of its water lost within the last 40 years. The reason is diversion of river water feeding the lake for agriculture, essentially cutting the lake off from its sole supply of water. As a result, the lake is disappearing and may be entirely gone by the year 2050. • Climate change and overexploitation are both blamed for the loss of over 90% of the water in Lake Chad (Africa).
The 'Ring of Fire':	The area surrounding the Pacific Ocean, from South America up to North America, over to Russia and back down through Southeast Asia, is known as the 'Ring of Fire'. Here, you will find the world's highest levels of earthquake and volcanic activity. Countries such as Japan and Indonesia are living under the constant daily threat of these geologic events.

SECTION I.2:   Review Questions

In this section you will review the main concepts and terminology from the Introductory Chapter by answering the questions provided below. Write in your answers in the space provided, and refer to the respective sections/pages of the chapter for the correct information.

Review Section I.2.1   The United States and the World (pp. 2–3)

1.) What is the difference between a 'state' and a 'State' in the world?

2.) The concept of a 'New World Order' would require what main actions? List at least three examples from the book:

A.)

B.)

C.)

Review Section I.2.2   A World on Maps (pp. 3–4)

1.) What is a 'mental map', and how is it commonly used by people?

2.) What is 'Cartography'? Give examples of how it has undergone a dramatic technological revolution:

3.) What are Geographic Information Systems, and how are they useful to Geographers?

Review Section I.2.3   Geography's Perspective (pp. 4–5)

1.) What is the Spatial Perspective, and how is it used in Geography?

2.) What role does environmental change have upon the spatial perspective of Geography?

3.) List some of the 'spatial vocabulary' terms that are commonly found in Geography:

Review Section I.2.4   Geography's Realms and Regions (pp. 5–10)

1.) What are Geographic Realms, in terms of their characteristics and properties?

2.) How is a Region different than a Realm in Geography? Give an example of this difference:

3.) List the four spatial classifications of the world, and briefly describe each in terms of size and distinct features:

   A.)

   B.)

   C.)

   D.)

4.) What are the three primary criteria for Geographic Realms? Briefly describe each.

5.) What are 'transition zones', and how do they relate to the concept of Geography? Give an example of a place in the world you will find such a zone.

6.) What are the two major categories that Geographic Realms can be divided into? Why?

7.) What is the 'Regional Concept'? What are the four main criteria to be a region? Briefly describe each.

8.) What are 'spatial systems'? How do the concepts of a functional region and hinterland relate to this overall idea?

Review Section I.2.5   The Physical Setting (pp. 10–17)

1.) List at least four examples of natural landscapes found throughout the world. In what ways could these be considered as barriers?

2.) Name at least two kinds of natural hazards you find on Earth. What is the 'Ring of Fire' and how does it apply to the idea of natural hazard location?

3.) In terms of climate, what are 'glaciations' and 'interglacials', and how have they affected the physical landscapes of Earth?

4.) How long ago was the last major glacial period in North America? When was it, and where in the USA was it primarily located?

5.) What is desertification? Give an example of where this is presently occurring in the world.

6.) What is the difference between the terms 'weather' and 'climate'?
   The next set of questions refers to the Köppen Climate System, found on pp. 15–17 of the text.
   Refer to both the textbook descriptions and Figure G – 7 (pp. 16–17) to assist you.

7.) What are A – Climates and where are they found? Give examples of three different types of A – Climates and their major characteristics, and list three different types of A – Climates.

8.) What are B – Climates and where are they found? Give examples of their major characteristics, and list two different types of B – Climates.

9.) What are C – Climates and where are they found? Give examples of their major characteristics.

10.) What are D– Climates and where are they found? Give examples of their major characteristics.

11.) What are E and H Climates? Give examples of their major respective characteristics.

Review Section I.2.6   Realms of Population (pp. 17–20)

1.) How do maps of population suggest the relative location of some of the world's realms?

2.) What is Earth's current population, and what percentage of the total available land area do humans occupy?

3.) Briefly describe each of the following world locations, in terms of their population density:

   A.) East Asia

   B.) South Asia

   C.) Europe

   D.) Eastern North America

4.) What is 'urbanization'? Where is it the highest and lowest in the world?

Review Section I.2.7   Realms of Culture (pp. 20–24)

1.) What is meant by the term 'Cultural Landscape'? How can this concept help define a realm?

2.) About how many languages are currently found in the world? What percentage of these are considered to be 'endangered'?

3.) What is a 'Language Family'? How many of these are found in the world?

4.) What is a 'Lingua Franca', and what language appears to be the primary example of this in the world?

Review Section I.2.8   Realms, Regions and States (pp. 25–27)

1.) What is the difference between the terms 'state' and 'nation – state'? What role does the European State Model have in these concepts?

2.) Across the world, states vary in several ways as seen in the subsection 'Power and the State' (p. 26). List at least five ways in which these differ worldwide.

   A.)

   B.)

   C.)

   D.)

   E.)

Review Section I.2.9   Patterns of Economic Development (pp. 27–33)

1.) Describe the field of Economic Geography, and how the concept of development is used to gauge a state in different ways.

2.) What are the four categories that countries are sorted into on the basis of their economic success?

3.) What are core and periphery areas, and what are the differences between these? Give an example of where you'd see this in the world.

4.) What is globalization? What are the pros and cons to globalization today? Give examples from the textbook to support your answer.

Review Section I.2.10   The Regional Framework (pp. 33–35)

1.) List and briefly describe the twelve realms of the world in terms of their location and major characteristics:

A.)

B.)

C.)

D.)

E.)

F.)

G.)

H.)

I.)

J.)

K.)

L.)

Review Section I.2.11   The Perspective of Geography (pp. 35–37)

1.) What are the concepts of Regional Geography and Systematic Geography, and how to they improve our study of world topics?

2.) What are some of the professional fields in which Geographers are employed?

SECTION I.3:   Blank Maps of The World

Blank maps are provided for you to utilize with the mapping questions from Section I.1 of the Study Guide. You are being provided with one blank map for each mapping exercise given in Section I.1. It is a good idea to make additional copies of these blank maps to use for extra practice and review.

SECTION I.4:   Student Companion Website

Additional study tools are available on the Student Companion Website at www.wiley.com/college/deblij. Features include:

- *Flashcards* offer an excellent way to practice key concepts, ideas, and terms from the text. You can review and quiz yourself on the concepts, ideas, and terms discussed in each chapter.

- *Map Quizzes* help you to master the place names for the various regions studies. Three game-formatted map activities are provided for each chapter.

- *Chapter Review Quizzes* provide immediate feedback to true/false, multiple choice, and short answer questions.

- *Audio Pronunciation* is provided for over 2000 key words and place names from the text.

- *Annotated Web Links*

- *Area and Demographic Data*

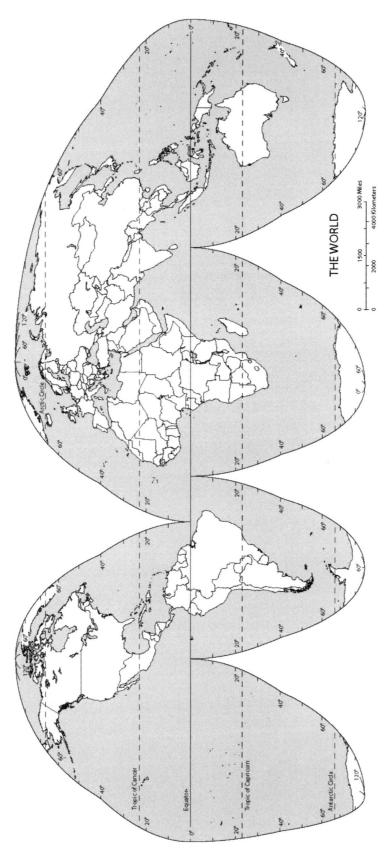

THE WORLD

3000 Miles
4000 Kilometers

Arctic Circle

Tropic of Cancer

Equator

Tropic of Capricorn

Antarctic Circle

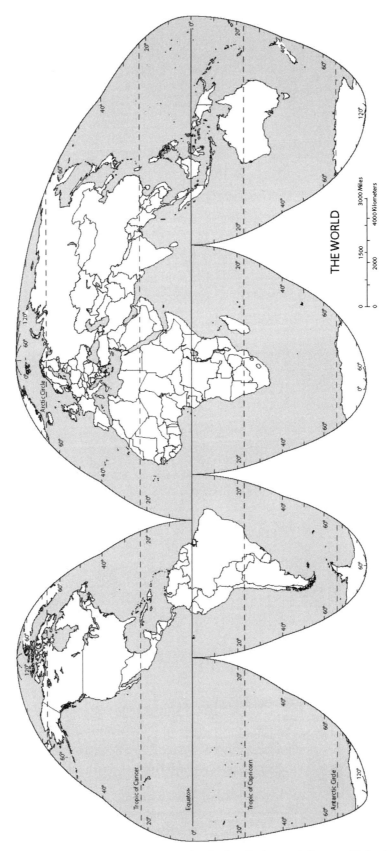

THE WORLD

3000 Miles

4000 Kilometers

Arctic Circle

Tropic of Cancer

Equator

Tropic of Capricorn

Antarctic Circle

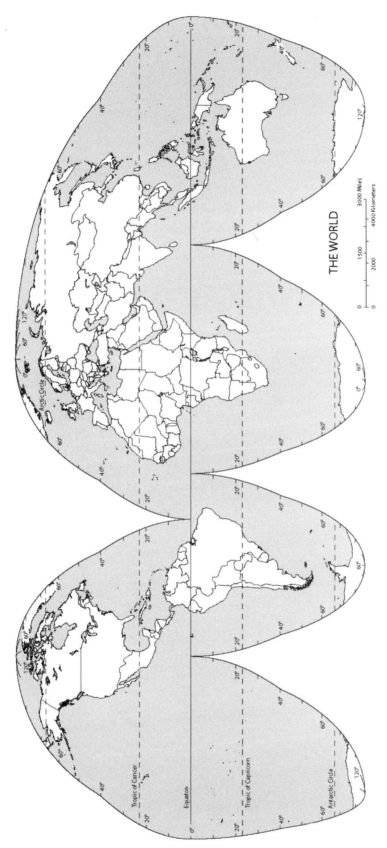

THE WORLD

3000 Miles
1500
4000 Kilometers
2000
0
0

Arctic Circle

Tropic of Cancer

Equator

Tropic of Capricorn

Antarctic Circle

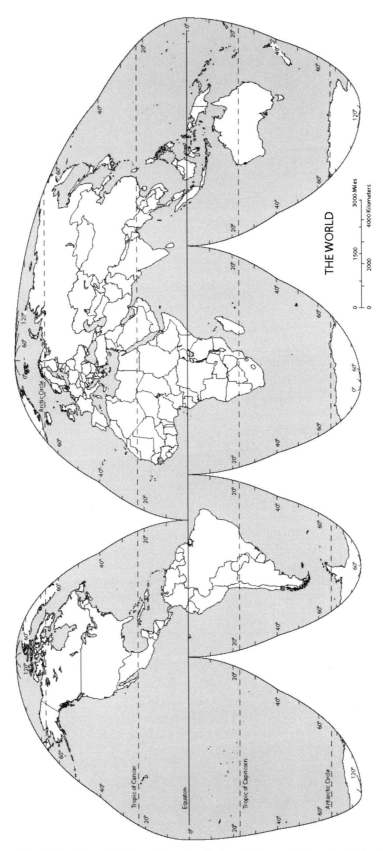

THE WORLD

3000 Miles
4000 Kilometers

1500

2000

0
0

Arctic Circle

Tropic of Cancer

Equator

Tropic of Capricorn

Antarctic Circle

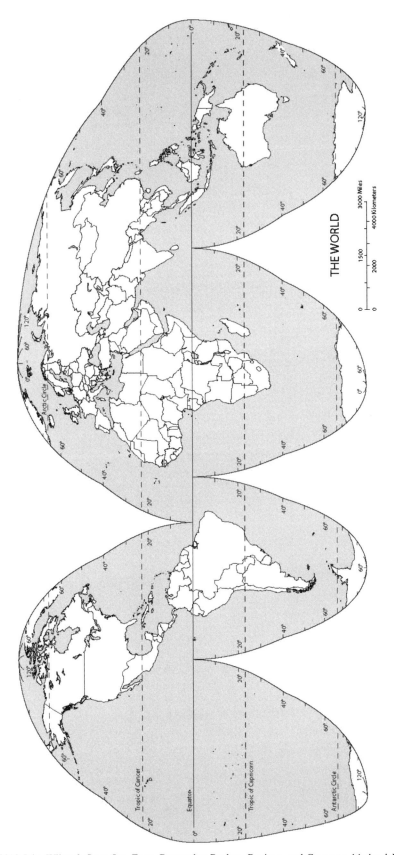

THE WORLD

| 0 | 1500 | 3000 Miles |

| 0 | 2000 | 4000 Kilometers |

Tropic of Cancer

Equator

Tropic of Capricorn

Antarctic Circle

Arctic Circle

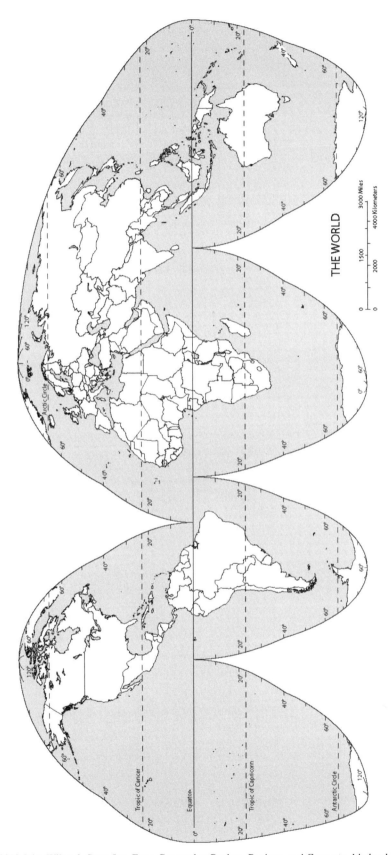

THE WORLD

3000 Miles
1500
0
4000 Kilometers
2000
0

Arctic Circle
Tropic of Cancer
Equator
Tropic of Capricorn
Antarctic Circle

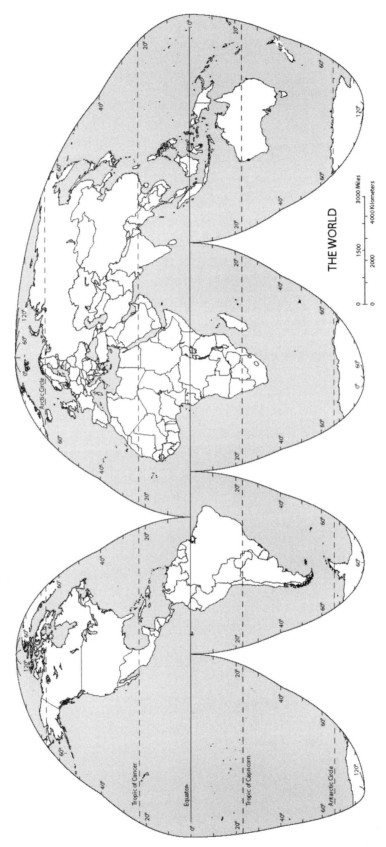

THE WORLD

3000 Miles
4000 Kilometers

1500
2000

0
0

Arctic Circle

Tropic of Cancer

Equator

Tropic of Capricorn

Antarctic Circle

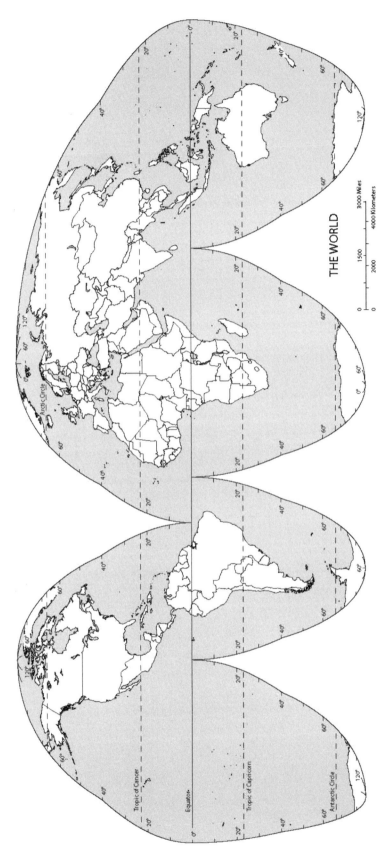

THE WORLD

	1500		3000 Miles
0			
0	2000	4000 Kilometers	

Arctic Circle

Tropic of Cancer

Equator

Tropic of Capricorn

Antarctic Circle

CHAPTER #1   EUROPE

SECTION 1.1:   Map Creation

In this section you will create your own detailed maps of the different major physical, cultural, geopolitical and environmental attributes of the realm. Each of these maps is critically important for understanding the Major Geographic Qualities of Europe, including their spatial distribution and functionality. By creating these on your own using the blank maps provided at the end of this chapter, you will actively learn where these different features are located and how their spatial distribution affects the ways of life in each of these respective areas.

Blank Maps for Europe are provided for you in Section 1.3. Label each of your maps with the appropriate name and title. For instance, the first map you will create will be called 'Map 1 – 1: Countries of Europe'.

Map 1–1:   Countries of Europe

There are 40 countries in the European Realm, including those in Eastern Europe that gained their independence due to the fall of the USSR in 1991. On the map, label the following countries of the European Realm. These have been divided into five regions for simplicity, and you may need to use abbreviations and/or arrows if you do not have enough space to write in the full name on the map. To assist with creation of this map, refer to the overall map of Europe found on pp. 38 - 39.

Hint: Write in the country name using a different colored pencil or pen for each of the regions these belong. For example, you could use a blue pencil for the Western Europe country names, a red pencil for the Eastern Europe country names, and so on.

Western Europe Countries:
- France   - Germany   - Switzerland   - Luxembourg
- Belgium   - Austria   - The Netherlands   - Lichtenstein

European Countries in the British Isles Region:
- The United Kingdom (U.K.)   - Ireland

Mediterranean Europe Countries:
- Spain   - Portugal   - Italy   - Greece   - Cyprus   - Malta

Northern European Countries:
- Iceland   - Norway   - Sweden   - Finland   - Denmark   - Estonia

Eastern European Countries:
- Latvia   - Lithuania   - Belarus   - Poland   - The Czech Republic   - Slovakia
- Ukraine   - Moldova   - Hungary   - Romania   - Montenegro   - Slovenia
- Croatia   - Bosnia   - Serbia   - Bulgaria   - Macedonia   - Albania

Map 1–2: Major Cities of Europe

On this map, label the following major cities of the European Realm. These have been divided into five regions as in Map 1-1, and you may need to use abbreviations and/or arrows if you do not have enough space to write in the full name on the map. To assist with creation of this map, refer to the overall map of Europe found on pp. 38 - 39. Write in the city name using a different colored pencil or pen for each of the regions these belong.

Major Western Europe Cities:
- Paris   - Berlin   - Munich   - Frankfort
- Amsterdam   - Brussels   - Lyon   - Vienna

Major European Cities in the British Isles Region:
- London   - Dublin   - Glasgow   - Liverpool

Major Mediterranean Europe Cities:
- Barcelona   - Lisbon   - Rome   - Venice   - Naples   - Madrid
- Marseille   - Sarajevo   - Nice   - Milan   - Athens   - San Marino

Major Northern European Cities:
- Oslo   - Stockholm   - Copenhagen   - Helsinki

Major Eastern European Cities:
- Prague   - Bratislava   - Budapest   - Bucharest
- Belgrade   - Warsaw   - Krakow   - Kiev

Map 1–3: Physical Features of Europe

Europe has a wide diversity of physical features throughout the realm that have played an important role in settlement patterns, political boundaries, distribution of mineral resources, and environmental issues as well. For this map, you will draw in the physical features listed below. These have been divided for you into categories, and it is recommended to use a different colored pencil or pen for each of the names per category.

To assist with creation of this map, refer to both the overall map of Europe found on pp. 38 – 39, and also Figure 1.4, p. 44.

Mountain Ranges: Using a blue pen/pencil, draw in the locations of these mountain ranges on your map and label them accordingly. Use a Λ symbol to denote the locations of these (i.e. draw a line of upside-down triangles to show where the mountain range belongs).
- Pyrenees Mountains   - Alps   - Scottish Highlands
- Carpathian Mountains   - Apennines   - Pindus Mountains

Peninsulas: A peninsula is defined as a land mass surrounded on three sides by water. Using a red colored pen/pencil, circle the following peninsulas on your map:
- Scandinavian Peninsula (Norway and Sweden)   - Jutland Peninsula (Denmark)
- Iberian Peninsula (Spain and Portugal)   - Italian Peninsula
- Balkan Peninsula (Slovenia, Greece, Romania and back)

General Physical Regions: These regions all have a physical characteristic in common, whether being lowlands or islands. Using Figure 1-4 (p. 52) as guide, circle and label the following physical regions on your map using a green pen/pencil:
- Northern European Lowlands   - Central Uplands   - Island of Corsica   - Balearic Islands
- Western Uplands   - Island of Sicily   - Island of Cyprus

Map 1–4: European Bodies of Water

Europe is a realm that has been heavily influenced by its water resources, both in terms of spatial history and also economic development. On this map, you will identify the major bodies of water found inside this realm. These are divided into groups and it is recommended to use a different colored pencil or pen for each of the names per water body category. Use pp. 38 – 39 as a guide.

Major Rivers:
- Rhine River   - Rhone River   - Po River   - Danube River   - Seine River   - Elbe River

Water Bodies in the Mediterranean Region:
- Mediterranean Sea   - Ionian Sea   - Adriatic Sea   - Balearic Sea   - Tyrrhenian Sea
- Strait of Gibraltar   - Black Sea   - Aegean Sea   - Strait of Sicily

Water Bodies by Northern Europe:
- Baltic Sea   - Gulf of Finland   - Gulf of Bothnia   - North Sea   - Barents Sea   - Norwegian Sea

Water Bodies by Western Europe and the British Isles:
- English Channel   - Celtic Sea   - Atlantic Ocean   - Bay of Biscay   - Irish Sea

Map 1–5: Major Climate Regions of Europe

Europe is a realm that contains a wide variety of climates. Though appearing to be complex when looking at the climate map in Figure 1-2 (p. 42) of the textbook, understanding the distribution of climates in this realm is not difficult if we break these into five categories:

A.) *Highland Climates* (those with mountains over 10,000 ft in elevation that contain a permanent snowcap throughout the year).

B.) *Mediterranean Climates* (those located primarily around the shores of the Mediterranean Sea, in Southern Europe).

C.) *Western European Climates* (Primarily C-climates located along/near the Atlantic Ocean, where the warmer ocean water has a moderating effect on the temperatures, allowing for milder winters).

D.) *Eastern European Climates* (Primarily D-climates in the interior of the European Realm, where these areas are dominated by a four-seasons climate due to their location between 40 – 70°N latitude).

E.) *Polar Climates*, found primarily above the Arctic Circle.

The major Köppen Climate types are given below for each of the categories above. Using a different colored pen/pencil for each category, outline and then color in these climate types:

Highland Climates:
- H – Climates

Polar Climates:
- E – Climates

Mediterranean Climates:
- Csa (The classic 'Mediterranean' Climate)
- BSk
- Csb

Eastern European Climates:
- Dfa
- Dfb
- Dfc

Western European Climates:
- Cfb
- Cfc

Map 1–6:   European Population Densities

The human populations across the European Realm are very unevenly distributed. Using the European Population Distribution map on Figure 1 – 6 (p. 48), we see that four general population density categories exist in the European Realm. Using a different colored pen/pencil for each category, outline and color the countries/sections of countries that fit into each of these designations. Once completed, you will see a spatial pattern of population distribution emerge across the realm.

Very Dense: Centered on Germany and France, extending into the southern half of the United Kingdom, Eastern Spain, and into Central Italy.

Dense: This population category accounts for the rest of Western Europe, along with all of Eastern and Mediterranean Europe.

Sparse: Southern and Central Scandinavia; some sparse populations are also found in the higher elevations of the Alps Mountain System.

Very Sparse: Primarily in the far northern parts of Scandinavia, along with northern Iceland.

Map 1–7:   Religions in Europe

The European Realm is home to significant religious diversity. Though these are primarily Christian religions, there are subgroups of Christianity that are spatially diverse and account for a range of cultural traditions and customs. Additionally, Islam is increasingly finding a presence in this realm. Although there is not a formal map of religions in the textbook, one can be created very simply in the following manner:

Part I: In the year 1054, the Christian Church split into Eastern (Orthodox) and Western (Catholic and Protestant) divisions. This split contains a spatial component, as it divided the European Realm into eastern and western sections. On your map, draw a bold-faced line on the border between the following countries, starting from the Mediterranean Sea and working north towards Finland:

The Eastern Boundary of the Line:
  - Greece, Bulgaria, Romania, The Ukraine, Belarus, Russia

The Western Boundary of the Line:
  - Albania, Macedonia, Serbia, Hungary, Slovakia, Poland, Lithuania, Latvia, Estonia, Finland

Part II: On your map, circle and color the following countries that belong to each category of religion listed below.

Eastern Orthodox Christianity:
  - Greece, Bulgaria, The Ukraine, Belarus, Russia

Main Catholic Countries:
  - Spain, France, Italy, Corsica, Hungary, Slovakia, Czech Republic, Poland, Lithuania, Latvia, Estonia, Finland, Ireland, Belgium

Main Protestant Countries:
  - Germany, Austria, the U.K., Denmark, Norway, Sweden, the Netherlands, Luxembourg

Countries with an Islamic Presence:
  - The former Yugoslavia Republics (Slovenia, Croatia, Serbia, Bosnia, Albania, Macedonia)

Map 1–8:   Languages of Europe

There are three different language families found in the European Realm. Of these, the Indo – European Language Family is the most dominant, accounting for over 95% of the European population. However, this family may be subdivided into Language Groups that include Germanic, Celtic, Romance, and more.

Using Figure 1-8 (p. 51), you will create a map of the primary locations of the Basque and Uralic Language Families, and then the subdivisions of the Indo – European Language Family. Use different colored pens/pencils for each category, and be sure to label these accordingly.

Basque Language Family: Located on the Northwestern border of Spain and France

Uralic Language Family: Finland, Estonia, and Hungary (Finland/Estonia area)

The Indo – European Language Family is divided into the following categories for this exercise:

Romance: Spain, France, Portugal, Italy, Romania, Belgium, Corsica

Germanic: Germany, Austria, The U.K., most of Ireland, Norway, Sweden, Iceland, Denmark, The Netherlands, Luxembourg

Slavic: Poland, Hungary, Czech Republic, Slovakia, Slovenia, Serbia, Croatia, Bosnia, Macedonia, Bulgaria, The Ukraine, Belarus

Celtic: Western Ireland, Northern U.K. (Scotland Region), Western U.K. (Wales)

Hellenic: Greece, Western Cyprus

Baltic: Latvia, Lithuania

Map 1–9:   Spatial History: The Evolution of the European Union

The European Union developed with three countries (Belgium, The Netherlands, and Luxembourg) that banded together in the midst of World War II, in an effort to strengthen their post-war status politically, socially, and economically. Over time, the concept of this union spread across Europe and forms the basis of the present-day European Union (EU).

However, not all of the realm is included in this Union. Southeast Europe and three former USSR Republics in Eastern Europe are not currently in the EU, due to civil and economic problems in their respective locations. Three other countries (Switzerland, Iceland, and Norway) actively voted against membership, and so their exclusion is intentional.

Using Figure 1-11 (p. 58), you will create a map that shows the spatial development of the European Union. As time progressed from World War II until present-day, the European Union developed spatially by starting in the center of the realm, expanding west, and then expanding north and then east.

Category 1:   *Original Members of the Union, by 1958:*
Category 2:   *Countries that Joined the Union between 1958 – 1990:*
Category 3:   *Countries Joining the Union between 1990 – 2007:*
Category 4:   *Countries voluntarily not in the Union today:*
Category 5:   *Countries either in discussion to become part of the Union, or not in the Union due to social, political, and/or economic issues:*

Map 1–10:   Geopolitical Issues in Europe

There are many Geopolitical Issues facing the European Realm today. Many of these have a spatial component, as certain groups of minorities want independence from the respective country in which they reside. Other issues stem from land rights to the continued growth and power of the European Union.

Using Figure 1-12 (p. 62) as a guide, circle and label the following areas on your map, displaying where these major Geopolitical Issues are found today. Use a different color pen/pencil (if possible) for each area, and consult the textbook for further information on the situation in these locations.

Part I: Prominent Geopolitical Issue Locations

Northern Irish: Found on the island of Ireland, this area belongs to the United Kingdom but is claimed by the Irish.

Scotland: Found in the northern region of the United Kingdom, this area is composed of people who want independence from the U.K.

Basques: This is an ethnic group found between the Spain and France border in the Pyrenees Mountains. These people want independence from Spain, and have used terrorism in the past to promote their efforts.

Alsaciens: These are German-speakers that live in the eastern region of France next to Germany. While not found on the textbook map, these people have a Geopolitical Issue in that they want fair representation from the French Government, who refuses to recognize them as a minority cultural entity.

Flemings and Walloons: These are two minority groups found in Belgium.

Corsicans: The Island of Corsica has been settled by both French and Italian people, both of whom want independence from the other on that land.

Catalans: These are a group of people totaling over 8 million that reside near Barcelona on the east coast of Spain. The Spanish Government recognizes this group and gives them political presence, though they want full independence.

Part II: Additional Geopolitical Issue Locations

Locate and circle the following places where European countries are dealing with issues of civil unrest, claims for independence, and/or border conflicts:
- *Kaliningrad (Russian Territory)* - *Upper Silesia* - *Crimea*
- *Kosovo (Newest Country of Europe)* - *Transylvania* - *Wales*

Map 1–11:   Environmental Issues in Europe

There are substantial environmental problems facing the European Realm today. This realm is one of the most advanced realms in the world today in terms of technology and emissions standards for air and water quality. However, moderate to severe environmental issues persist in certain locations.

These environmental issues are divided into categories below. Using a different colored pen/pencil for each, circle or highlight the areas and label them using their respective environmental situation.

Sealevel Rise:	The lowland regions of Western Europe are the most susceptible to sealevel rise, particularly The Netherlands, Northern France, and Northern Germany.
Coastal Pollution:	Almost all of the Mediterranean Sea along Europe's southern boundary is affected by coastal pollution, sources primarily being urban sewage and manufacturing waste that are discharged directly into this body of water.
	Additionally, the English Channel, North Sea and Baltic Sea have problems with coastal pollution for similar reasons.
River Pollution:	Most of Europe's major rivers have some form of water quality issues. In particular, the Rhine, Rhone, Danube (From Austria to The Ukraine), and most of the rivers in Germany and Poland are affected. Trace these using a blue pen/ pencil and label them accordingly.
Air Quality Issues:	As a result of the heavy manufacturing and industry presence in the European Realm along with heavy urban population concentrations, poor urban air quality has been an issue in the realm for some time. Acid Rain also continues to be a problem in these regions. Draw a line around the worst areas of air quality, using the steps below as your guide:
	Start the line in Central France → Southern Germany → The Western Border of the Ukraine → North to Central Finland → West to Central Norway → West to the U.K. (excluding Scotland and Ireland) → back down to Northern France.

Map 1–12: *Regions of the European Realm*

Using Figure 1-13 (p. 65) in the textbook, color the following regions of the European Realm, using a different colored pen/pencil for each region and the country list below as a guide. After completing this exercise, consult the textbook to understand why each of these is considered a separate region of the realm.

Western Europe:

- France	- Germany	- Switzerland	- Luxembourg
- Belgium	- Austria	- The Netherlands	- Lichtenstein

The British Isles: - The United Kingdom (U.K.)   - Ireland

Mediterranean Europe:
- Spain   - Portugal   - Italy   - Greece   - Cyprus   - Malta

Northern Europe:
- Iceland   - Norway   - Sweden   - Finland   - Denmark   - Estonia

Eastern European:

- Latvia	- Lithuania	- Belarus	- Poland	- The Czech Republic	- Slovakia
- Ukraine	- Moldova	- Hungary	- Romania	- Montenegro	- Slovenia
- Croatia	- Bosnia	- Serbia	- Bulgaria	- Macedonia	- Albania

SECTION 1.2: Review Questions

In this section you will review the main concepts and terminology from the European Realm by answering the questions provided below. Write in your answers in the space provided, and refer to the respective sections/pages of the chapter for the correct information.

Review Section 1.2.1   Europe's Major Geographic Qualities (p. 41)

Summarize the eleven Major Geographic Qualities of Europe in your own words:

1.)

2.)

3.)

4.)

5.)

6.)

7.)

8.)

9.)

10.)

11.)

Review Section 1.2.2   Geographical Features (pp. 41–43)

1.) There are how many countries in the European Realm?

2.) What is the estimated overall population?

3.) For many centuries, Europe has been a hearth of _____, _____, and _____.

4.) Where is the Eastern boundary of the European Realm, according to our regional definition?

5.) What are some natural resources contained in the European realm that have benefitted its people, both in historical and recent times?

6.) Why has the cultural diversity of Europe been both an asset and a liability?

7.) Name some the reasons that Europe would be considered to have an outstanding locational advantage:

Review Section 1.2.3   Landscapes and Opportunities (p. 43)

1.) What does *physiography* mean? What are some of its components?

2.) Name and briefly describe the four physiographic units of Europe below:

Review Section 1.2.4   Historical Geography (pp. 43–45)

1.) In what two countries did Europe witness the rise of its first great civilizations?

2.) What contributions did the Greeks make to the foundations of European civilization?

3.) Where was the Roman Empire located, relative to present-day Europe?

4.) What does the term *locational functional specialization* mean? Give some examples of locations and their specialties that the Romans exploited.

5.) When did the Roman Empire collapse? What groups of people migrated into the former Roman territory, and where did they locate?

6.) When was the European Renaissance?

7.) What role did *mercantilism* have on Europe and its Renaissance?

Review Section 1.2.5   The Revolutions of Modernizing Europe (pp. 45 – 47)

1.) What was the *Agrarian Revolution*? How did this impact land ownership and agricultural methods?

2.) What was the central idea of Von Thünen's Model? How did this produce the idea of *location theory*?

3.) What was the *Industrial Revolution*? Which European countries benefitted the most from it?

4.) What mineral deposits in Europe helped propel the Industrial Revolution? Where are they primarily found?

5.) What are the main ideas of Alfred Weber's industrial location model? How is this different from Von Thünen's Model?

6.) What does Europe's *Political Revolution* refer to, and when did this occur?

7.) What does the term *nation-state* mean? How is it different from the idea of a *nation*?

8.) What are *centripetal forces* and *centrifugal forces*? Give an example of how each has affected Europe in recent times.

Review Section 1.2.6   Contemporary Europe (pp. 50 – 56)

1.) What aspects of Europe would qualify it as being a 'modern' landscape of human activity? List at least four different reasons.

2.) What is the primary language family spoken in Europe? Give examples.

3.) What is a *Lingua Franca*?

4.) What is the primary religion practiced in Europe? Are there any minority religions, and if so what are they?

5.) Europe is unified by three major principles, according to Edward Ullman. List and briefly describe these three in your own words:

6.) What is a *Primate City*? Give an example of one you will find in Europe.

7.) What do the terms *metropolitan area* and *Central Business District* mean?

8.) Why will Europe possibly experience a *population implosion* in this century?

9.) What countries in Europe are currently experiencing strong immigration into their borders? Where are these immigrants coming from?

Review Section 1.2.7   Europe's Modern Transformation (pp. 56 – 63)

1.) Who proposed the recovery program for Europe that was instituted after World War II? Briefly describe the details of this program.

2.) What was the Benelux Agreement? When was it signed, and by what countries?

3.) What is *Supranationalism*?

4.) Starting from 1957 through 2007, name the countries that joined the present-day European Union and the dates they were admitted into the EU.

5.) What is the *Euro*?

6.) What two groups today are not part of the European Union? Why?

7.) Why does the EU want to admit Turkey into the Union? What are the potential problems with this?

8.) What is meant by the term *Devolution*? Give examples of how this has affected the European Realm.

9.) What steps have European planners taken to reduce the divisive effects of their national borders?

10.) What are the *Four Motors of Europe*? Where are they located?

Review Section 1.2.8   Regions of the Realm: Core and Periphery (pp. 63–64)

1.) Why does Europe present a challenge in terms of categorizing it by regions?

2.) What are the four largest cities in Europe, in terms of population?

3.) What are *Microstates*? Give at least four examples from Europe.

4.) Why has the old regionalization scheme of Europe become outdated?

Review Section 1.2.9   States of the Mainland Core (pp. 64–72)

1.) What eight countries constitute the 'Core of Europe'?

2.) Describe the differences between East Germany and West Germany, between the end of World War II and the fall of the 'Iron Curtain'.

3.) Name the current 'States' of reunified Germany. How are these different between the western and eastern parts of the country?

4.) Compare France versus Germany, and why these two countries are vastly different.

5.) What do the concepts of *site* and *situation* mean, and how are they applied to the city of Paris, France?

6.) Briefly describe France's economic geography attributes today.

7.) What impact did Napoleon have on the regional geography of France?

8.) Why has the area around Lyon, France, become an 'economic powerhouse'?

9.) What three countries constitute Benelux?

10.) What three countries comprise the 'Alpine States'? Briefly mention their contributions to Europe, whether economically, physically, and/or politically.

11.) In what year did the Czech Republic and Slovakia become independent of each other?

Review Section 1.2.10   The Core Offshore: The British Isles (pp. 72 – 77)

1.) What two countries constitute the 'British Isles'? What significance does the year 1921 have upon these two countries?

2.) The 1707 'Act of Union' resulted in the United Kingdom being created from which four countries?

3.) What impact did the Industrial Revolution have on Britain?

4.) What are the four present-day subregions of the United Kingdom? On what basis were they divided into these?

5.) What factors in the Irish Republic have combined to produce the highest rate of economic growth in the entire European Union?

6.) Why had Ireland's first economic boom fade by the year 2009?

Review Section 1.2.11   The Contiguous Core in the South (pp. 77 – 82)

1.) What four countries constitute this region?

2.) Why is Italy often described as being 'two countries'? What are the names of these two subregions, and how are they different?

3.) What is the *Ancona Line*, and what is its significance?

4.) What part of Italy receives the greatest number of illegal immigrants? What effect do these people have on this part of the country?

5.) Where is the *Iberian Peninsula*? What physical features isolate this region from the rest of Europe?

6.) What effect have devolutionary pressures had upon Spain in recent times?

7.) Where is 'Catalonia', and what economic impact does it have on Spain?

8.) How has the country of Portugal benefitted from joining the EU?

9.) Where is the country of Malta located? When did it join the EU?

Review Section 1.2.12   The Discontinuous Core in the North (pp. 82 – 85)

1.) What six countries comprise this region? What is their combined population?

2.) How have remoteness, isolation and severe environmental conditions actually helped this region over time?

3.) What are the three major languages in this region?

4.) Give some examples of modern Swedish manufacturing today:

5.) What is a *break-of-bulk* function? Which country in this realm is a prime example of this idea, and why?

6.) The former Denmark territory of Greenland is today is known by what name?

7.) What natural resources sustain Finland's economy?

8.) Why is the country of Estonia considered to be part of this region?

9.) Briefly describe the economic geography of Iceland:

Review Section 1.2.13   The Eastern Periphery (pp. 85 – 97)

1.) What do the terms *Shatter Belt* and *Balkanization* mean? Where will you find such examples in this region of Europe?

2.) What is *Ethnic Cleansing*, and how has it impacted this region?

3.) What four groups of countries have been created from this region?

4.) What is Kaliningrad, and where is it located?

5.) What countries are a part of the 'EU Countries Contiguous to the European Core'?

6.) What effect did joining the EU have upon Poland?

7.) Who are the Roma? Briefly describe their effect on this region:

8.) What country in this region has become synonymous with corruption and inefficiency? What is its capital city?

9.) What does the concept of *Irredentism* mean? What country in this region has primarily experienced this phenomenon?

10.) Which was the first 'Republic' of former Yugoslavia to be invited into the EU?

11.) About how many islands comprise the Greek Archipelago?

12.) How has Greece been affected by the recent enlargement of the EU?

13.) Where is the island country of Cyprus located? What two ethnic groups are claiming this land?

14.) Why do most Europeans feel that Romania has no place in the EU?

15.) What is Bulgaria's main economic advantage?

16.) What primary problem prevents Latvia and Lithuania from becoming a part of the European Core?

17.) What are the seven countries that gained independence as a result of the breakup of Yugoslavia? (Hint: Figure 1-26, p. 94)

18.) What countries in this region are primarily Muslim in religion?

19.) Why is the Ukraine's location crucial for the European Realm?

20.) How does the Dnieper River geographically divide the Ukraine?

21.) Why is the economy of Moldova considered to be on the decline today?

22.) The country of Belarus is considered to be the most 'backward' of all in the European Realm for what reasons?

SECTION 1.3:   Blank Maps of Europe

Blank maps are provided for you to utilize with the mapping questions from Section 1.1 of the Study Guide. You are being provided with one blank map for each mapping exercise given in Section 1.1. It is a good idea to make additional copies of these blank maps to use for extra practice and review.

SECTION 1.4:   Student Companion Website

Additional study tools are available on the Student Companion Website at www.wiley.com/college/deblij. Features include:

- *Flashcards* offer an excellent way to practice key concepts, ideas, and terms from the text. You can review and quiz yourself on the concepts, ideas, and terms discussed in each chapter.
- *Map Quizzes* help you to master the place names for the various regions studies. Three game-formatted map activities are provided for each chapter.
- *Chapter Review Quizzes* provide immediate feedback to true/false, multiple choice, and short answer questions.
- *Audio Pronunciation* is provided for over 2000 key words and place names from the text.
- *Annotated Web Links*
- *Area and Demographic Data*

EUROPE

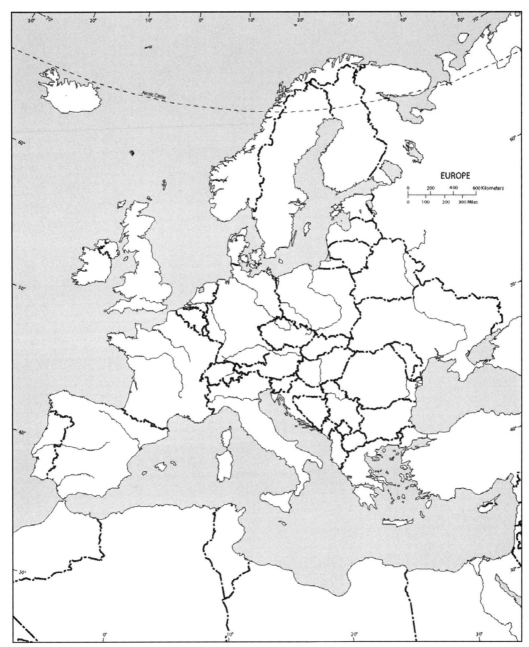

EUROPE

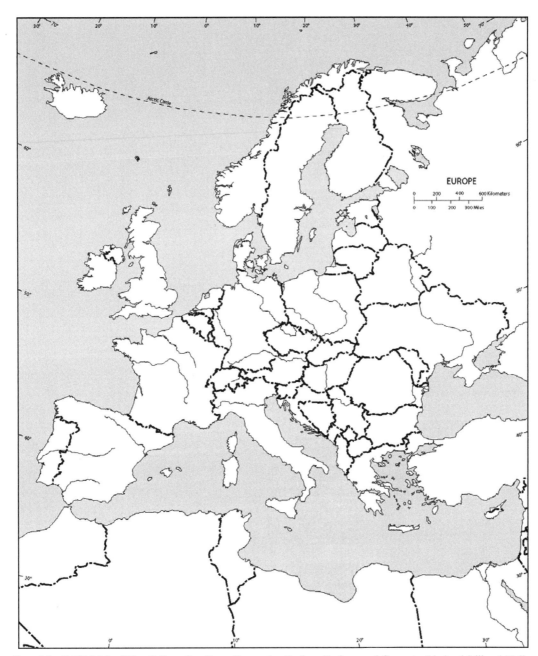

EUROPE

| 0 | 200 | 400 | 600 Kilometers |
| 0 | 100 | 200 | 300 Miles |

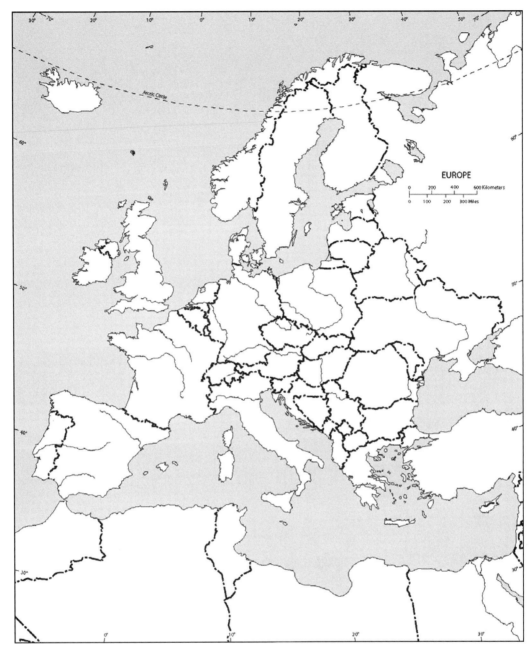

EUROPE

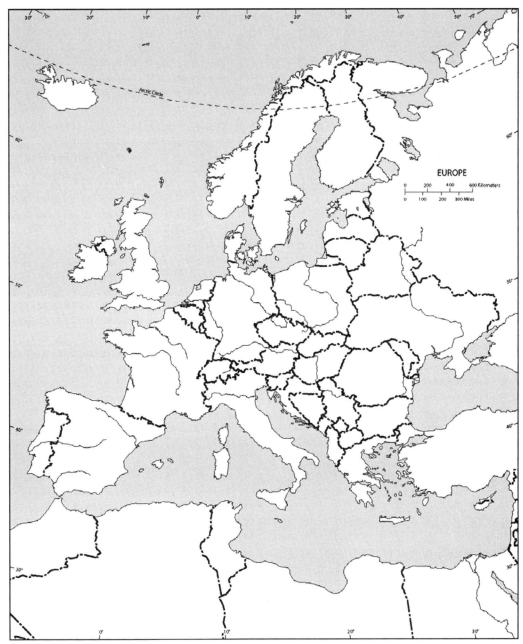

EUROPE

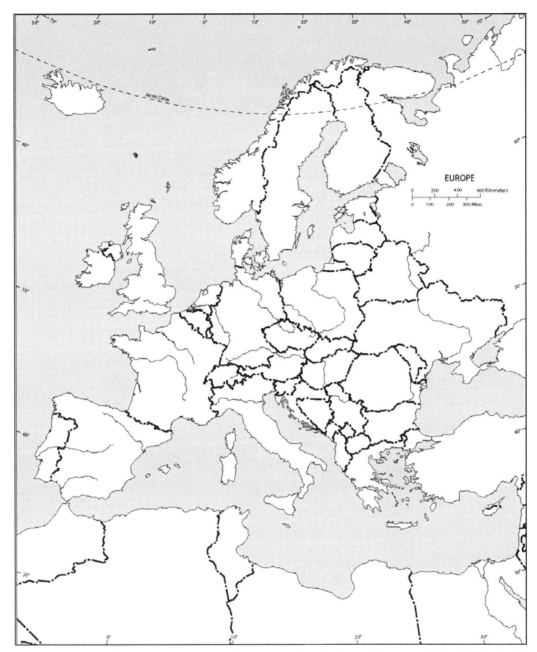

EUROPE

0   200   400   600 Kilometers
0   100   200   300 Miles

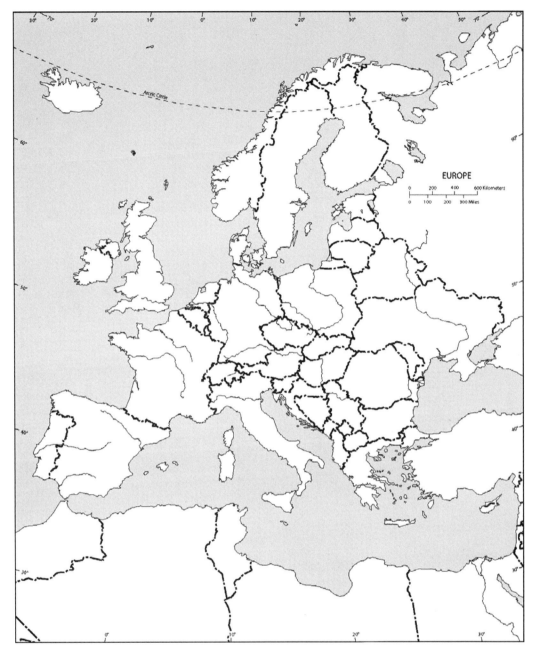

EUROPE

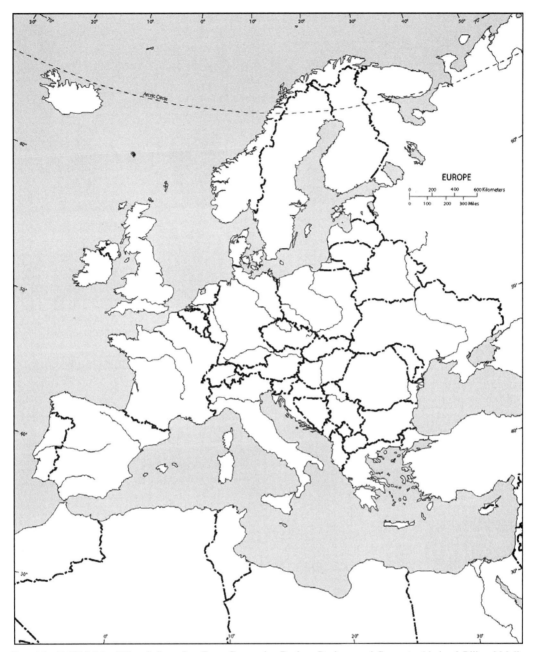

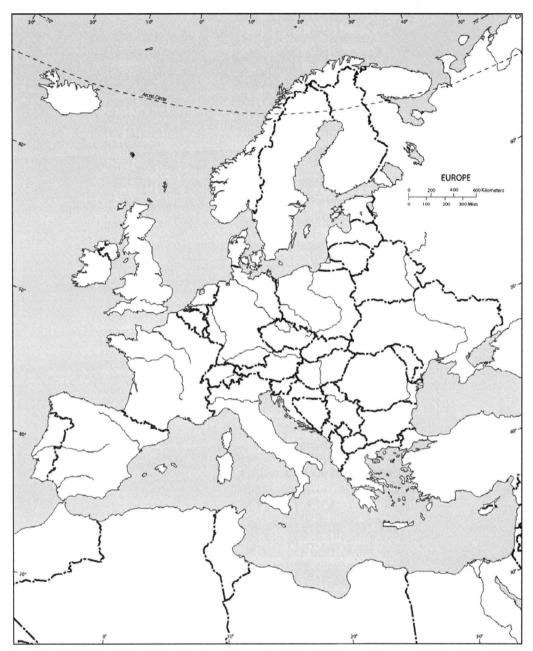

EUROPE

0   200   400   600 Kilometers
0   100   200   300 Miles

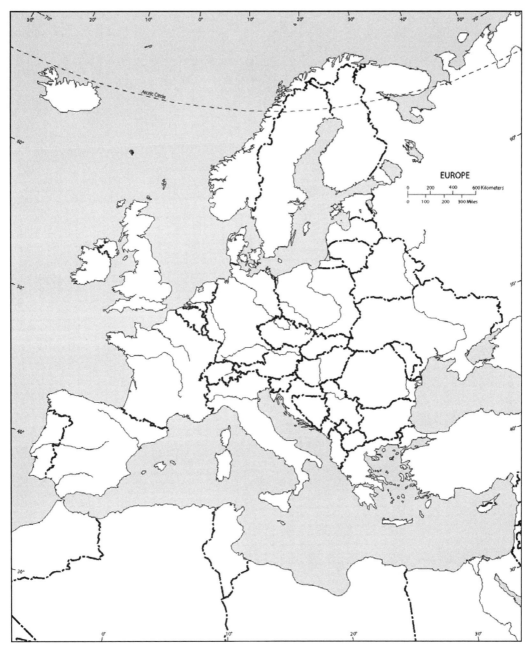

EUROPE

0   200   400   600 Kilometers
0   100   200   300 Miles

CHAPTER #2   RUSSIA

SECTION 2.1: Map Creation

In this section you will create your own detailed maps of the different major physical, cultural, geopolitical and environmental attributes of the realm. Each of these maps is critically important for understanding the Major Geographic Qualities of Russia, including their spatial distribution and functionality. By creating these on your own using the blank maps provided at the end of this chapter, you will actively learn where these different features are located and how their spatial distribution affects the ways of life in each of these respective areas.

Blank Maps for Russia are provided for you in Section 2.3. Label each of your maps with the appropriate name and title. For instance, the first map you will create will be called 'Map 2–1: Former States of the Soviet Union'.

Map 2–1: Former States of the Soviet Union

There are 15 former states of the Soviet Union that gained their independence due to the fall of the U.S.S.R. in 1991, with Russia being the largest.

On the map, label the following countries of the Russian Realm. These have been divided into three regions, and you may need to use abbreviations and/or arrows if you do not have enough space to write in the full name on the map. To assist with creation of this map, refer to Figure 2–7, p. 117 (The Former Soviet Empire).

Former USSR States in the West:
- Lithuania   - Estonia   - The Ukraine
- Latvia   - Belarus   - Moldova

Former USSR States in the South:
- Kazakhstan   - Turkestan   - Uzbekistan
- Kyrgyzstan   - Tajikistan

Former USSR States in the Southeast:
- Georgia   - Armenia   - Azerbaijan

Map 2–2:   Major Cities of Russia

On this map, label the following major cities of the Russian Realm. You may need to use abbreviations and/or arrows if you do not have enough space to write in the full name on the map. To assist with creation of this map, refer to the overall map of Russia found on pp. 98 - 99. Write in the city name using a different colored pencil or pen for each of the regions these belong.

Major Western Russia Cities:

- Moscow	- St. Petersburg	- Volgograd	- Rostov
- Samara	- Murmansk	- Perm	- Chelyabinsk
- Bryansk	- Omsk	- Novosibirsk	- Astrakhan

Major Eastern Russia Cities:

- Irkutsk	- Bratsk	- Vladivostok	- Khabarovsk
- Yakutsk	- Komsomolsk		

Map 2–3:   Physical Features of Russia

Russia has a wide diversity of physical features throughout the realm that have played an important role in settlement patterns, political boundaries, distribution of mineral resources, and environmental issues as well. For this map, you will draw in the physical features listed below. These have been divided for you into categories, and it is recommended to use a different colored pencil or pen for each of the names per category. To assist with creation of this map, refer to both the overall map of Russia found on pp. 98–99, and also Figure 2–2, p. 106.

Mountain Ranges: Using a blue pen/pencil, draw in the locations of these mountain ranges on your map and label them accordingly. Use a Λ symbol to denote the locations of these (i.e. draw a line of upside-down triangles to show where the mountain range belongs).

- Ural Mountains	- Caucasus Mountains	- Sayan Mountains
- Verkhoyansk Range	- Cherisky Range	- Kolyma Range
- Koryak Range	- Dzhugdzhur Range	- Stanoyoy Range
- Yablonovyy Range	- Sikhote – Alin Range	

Peninsulas: A peninsula is defined as a land mass surrounded on three sides by water. Using a red colored pen/pencil, circle the following peninsulas on your map:

- Yamal Peninsula	- Gyda Peninsula	- Taymyr Peninsula
- Kamchakta Peninsula	- Kola Peninsula	- Chukchi Peninsula

Major Islands: These islands are part of Russian Territory, and can be strategic (i.e. military bases and/ or weapon placement), or have significant mineral deposits. Circle and name these islands below:

- Kurile Islands	- Commander Islands	- Sakhalin Island
- New Siberian Islands	- Shantar Islands	- Novaya Zemlya Island
- Belyy Island	- Bolshevik Island	- October Revolution Island

Map 2–4: Major Physical Regions of Russia

Based on the wide variety of Russia's physical features, geographers have divided the realm into several physiographic regions. These contain similar physical features within their respective boundaries, from mountains to plateaus. You will create a map and draw/label where these physiographic regions are located, using Figure 2–2, p. 106 as your guide.

- The Russian Core
- The Ural Mountains
- The West Siberian Plain
- The Central Siberian Plateau

- The Eastern Highlands
- The Central Asian Ranges
- The Caucasus Mountains
- The Yakutsk Basin

Map 2–5: Russian Bodies of Water

Russia is a realm that has been influenced by its water resources, both in terms of spatial history and also economic development. On this map, you will identify the major bodies of water found inside this realm. These are divided into groups and it is recommended to use a different colored pencil or pen for each of the names per water body category. Use pp. 98–99 as a guide.

Major Rivers:
- Volga River
- Lena River

- Dama River
- Aldan River

- Ob River
- Angara River

- Yenisey River
- Kolyma River

Inland Bodies of Water:
- Black Sea
- Lake Ladoga

- Caspian Sea
- Lake Khanka

- Lake Baikal

- Lake Onega

Saltwater Seas and Oceans:
- Arctic Ocean
- Gulf of Ob
- Gulf of Anadyr

- Pacific Ocean
- White Sea
- Sea of Okhotsk

- Kara Sea
- Laptev Sea
- Tatar Strait

- Bering Sea
- East Siberian Sea
- Bering Strait

Map 2–6: Major Climate Regions of Russia

Russia is a realm that contains a wide variety of climates. Most of these climates contain very cold temperatures throughout the year, although there are some slightly warmer climates found in the extreme southern portions of the realm. Use the World Climate Map (Figure G – 7, pp. 16–17) as a guide.

The major Köppen Climate types are given below for each of the climates found in this realm. Using a different colored pen/pencil for each category, outline and then color in these climate types:

Highland Climates:
- H – Climates

Polar Climates:
- E – Climates

Cold Mid – Latitude Climates:
- Dfa
- Dfb
- Dfc
- Dfd

- Dwb
- Dwc
- Dwd

Dry Climates:
- BSk

Map 2–7: ***Russian Population Densities***

The human populations across the Russian Realm are some of the most sparse in the world. They are also very spatially uneven, as Western Russia contains over 80% of the population of the realm. Using the Russian Population Distribution map on Figure 2–4 (p. 108), we see that four general population density categories exist in the Russian Realm. Using a different colored pen/pencil for each category, outline and color the countries/sections of countries that fit into each of these designations. Once completed, you will see a spatial pattern of population distribution emerge across the realm.

Dense: This population category is found only Western Russia, from Moscow to the Ural Mountains and south to the Caucasus Mountains.

Sparse: This population density is found in the rest of Western Russia, to just east of the Ural Mountains. It is also found centered on Vladivostok in extreme Southeast Russia.

Very Sparse: This is the dominant population density category in all of Eastern and Southern Russia, with the exception of the northern regions including Siberia.

Empty: Along the northern fringe of Russia and also containing most of NE Russia (Siberia).

Map 2–8: ***Spatial History: The Evolution of the Russian Union***

Using Figure 2–5 (p. 110), you will create a map that shows the spatial development of the Russian Union. As time progressed from the 1500's until present-day, the Russian Union developed spatially by starting in the western part of the present-day realm, expanding east to the Pacific Ocean. Using a different colored pen/pencil, draw in the periods of spatial acquisition as shown in the categories below.

Category 1: Russian territory acquired by the year 1533

Category 2: Russian lands obtained from 1533–1598

Category 3: Regions acquired by Russia from 1598–1689

Category 4: Land acquired by Russia from 1689–1725

Category 5: Land annexed to Russia between 1725–1801

Category 6: Regions obtained between 1801–1945

Map 2–9:   The New Federal Districts of Russia and Population Decline

Using Figure 2–10 (p. 124), you will create a map that shows the present – day Federal Districts of Russia, their respective capitals, and the percentage of population decline since 1991. Use a different colored pen/pencil for each district:

Name of District	*District Capital City*	*Percentage of Population Decline*
Northwest	St. Petersburg	−9%
Central	Moscow	−0.2%
South	Rostov	−12%
Volga	Nizhniy Novgorod	−2%
Ural	Yekaterinburg	−2%
Siberia	Novosibirsk	−5%
Far East	Khabarovsk	−17%

Map 2–10:   Environmental Issues in Russia

The Russian Realm is one of the most environmentally-degraded places in the world today. Since the fall of the USSR in 1991, environmental regulations have been scarce at best, allowing contamination of almost every major body of water in the entire realm. In addition, mining has led to severely contaminated lands- some of which may never be able to regenerate, according to scientists.

These environmental issues are divided into categories below. Using a different colored pen/pencil for each, circle or highlight the areas and label them using their respective environmental situation.

Lake Baikal: Lake Baikal is over 5,000 feet deep and is estimated to contain over 20% of the world's accessible freshwater supply. This is staggering, considering almost all of the lake is heavily polluted- one paper alone discharged over 1 BILLION gallons of pollution into the lake in the 1960's! Bear in mind that one gallon of pollution will contaminate over 1 million gallons of fresh water! Combined with many other factories and agricultural runoff, and Lake Baikal is one of the most contaminated lakes on the planet.

Coastal Pollution: Almost all of the Arctic Ocean along Russia's northern boundary is affected by coastal pollution. The sources are manufacturing waste that are discharged directly into this body of water, and direct dumping of chemical and nuclear waste barrels onto the beaches adjacent to the ocean.

Additionally, the Pacific Ocean coastline centered on Vladivostok, contains many problems with contaminated seawater for similar reasons, along with oil and gas spills from tankers that frequently use this port.

River Pollution: Almost all of Russia's major rivers have some form of water quality issues. In particular, the Volga, Ob, Yenisey, and Lena Rivers all have moderate to sever contamination. Trace these using a blue pen/pencil and label them accordingly.

Air Quality Issues:	As a result of the heavy manufacturing and industry presence in the Russian Realm along with little to no environmental regulations, air quality has been a substantial issue in the Russian Realm since the rise of manufacturing in the 1800's. In particular, the regions from Moscow to St. Petersburg and extending to the Ural Mountains contain the worst air pollution levels overall in the realm.
Deforestation:	Regions of Northwest Russia, crossing the Ural Mountains and stretching to Lake Baikal, have been heavily deforested in the last 100 years. Most of these areas have not seen reforestation efforts, and combined with toxic soils and groundwater levels nearby, these forests could take generations- if ever- to become replenished.

Map 2–11:   Regions of the Russian Realm

Using Figure 2–11 (p. 129) in the textbook, color the following regions of the Russian Realm and use a different colored pen/pencil for each region and the country list below as a guide. After completing this exercise, consult the textbook to understand why each of these is considered a separate region of the realm.

- Russian Core   - Siberia   - Eastern Frontier   - Far East

SECTION 2.2: Review Questions

In this section you will review the main concepts and terminology from the Russian Realm by answering the questions provided below. Write in your answers in the space provided, and refer to the respective sections/pages of the chapter for the correct information.

Review Section 2.2.1   Russia's Major Geographic Qualities (p. 102)

Summarize the ten Major Geographic Qualities of Russia in your own words:
  1.)
  2.)
  3.)
  4.)
  5.)
  6.)
  7.)
  8.)
  9.)
 10.)

Review Section 2.2.2   Introduction to the Realm (pp. 101–102)

1.) What is the current population of Russia today? Is it increasing, decreasing, or staying constant?

2.) After the Soviet Union collapsed in 1991, it broke into Russia and 14 newly independent countries. List these countries, and their respective populations:

3.) Who are the 'Oligarchs'? What impact have they had on Russia in present times?

4.) Who are currently the two most influential political men in Russia today? What are their roles?

5.) Describe Transcaucasia in terms of location and current geopolitical conflicts in this realm:

Review Section 2.2.3   Roots of the Russian Realm (pp. 103–105)

1.) Describe Precommunist Russia's culture, and how the Industrial Revolution caused some modification to this culture.

2.) What were the primary driving forces that sent Russian armies expand eastward across Siberia and to the Pacific Ocean?

3.) What happened in 1917 that led to the beginning of the Soviet Union?

4.) Though Russia was dominant in the new Soviet Union, its name disappeared from the map as it merged with other republics. What role did the czars have for these other republics?

5.) When did the USSR exist as a political entity?

Review Section 2.2.4   Russia's Physical Geography (pp. 105–109)

1.) What features dominate Russia's physical geography across the realm?

2.) Describe the types of climate you will find in Russia both in summer and winter, and how this affected both settlement and agriculture throughout the realm.

3.) What is the field of 'Climatology'? How does the 'continentality' of Russia influence its climate?

4.) How is the Arctic environment changing, and what impacts could this have on Russia?

5.) Though the western regions of Russia are the most climatically favorable for agriculture, what problems exist that make farming difficult?

6.) What are the 'Taiga' and 'Tundra', where are they found in the realm and how are they different?

7.) Name the major mountain ranges in the Russian Realm, and where you will find each.

8.) Describe Siberia in terms of location, physical landscapes and climate:

Review Section 2.2.5   Evolution of the Russian State (pp. 109–112)

1.) What is a 'Rus', and what impact did it have on early development of the Russian Realm?

2.) Briefly describe the spatial development of the present-day Russian Realm from up until the 1700's, explaining how the Tatar and Cossack peoples influenced this expansion.

3.) What impacts did Czar Peter the Great and Czarina Catherine the Great have upon spatial expansion of the Russian empire?

4.) Why were the Russians preoccupied with conquering Central Asia in the 1800's?

5.) What impact did the Russo – Japanese War of 1904–1905 have upon both the spatial and social development of Russia?

Review Section 2.2.6   The Colonial and Soviet Legacies (pp. 112–120)

1.) How large (in terms of land area) was the Russian Empire at its greatest extent at the start of the 1900's? What role did 'imperialism' have upon the continued expansion of Russia?

2.) Why was the idea of communism so popular in Russia in the 1910's and 1920's? Give specific examples from the textbook to support your answer.

3.) Who were the Bolsheviks and Mensheviks, and what role did these groups play in the Russian Revolution? What city was the national capital moved to in 1914 as a result?

4.) What is the concept of 'Russification', and how did it spatially separate cultures in the realm?

5.) What was the intended purpose of the 'Federations' in the Soviet Union? Where were they located?

6.) Soviet planners had what two primary objectives when instituting communism in the 1920's? How does the concept of a 'sovkhoz' relate to these ideas?

7.) What is meant by the term 'Command Economy', and how did the Soviet Union practice this? Give specific examples from the textbook to support your answer.

Review Section 2.2.7   Russia's Changing Political Geography (pp. 120–124)

1.) What was the goal of the 'Russian Federation Treaty' in 1992, and what spatial impact was it intended to have upon the realm?

2.) Where is Chechnya, who are the people that inhabit this territory, and what geopolitical conflicts have they had with Russia in the 1990's?

3.) Describe the concept of a 'Federal State System'. Is this type of government currently being practiced in Russia, why or why not?

4.) How do Russia's physical size and the concept of 'distance decay' impact the realm in terms of modern-day development?

5.) Who was elected as President of Russia in 2000? What impacts did he have upon the realm?

6.) The New Era of Globalization in Russia is characterized by what kinds of freedoms?

7.) List the five main geopolitical issues found in the realm today, starting with 'The Revival of Religion'. Briefly describe each and the impact they are having upon the realm's way of life.

A.)

B.)

C.)

D.)

E.)

Review Section 2.2.8   Russia's Demographic Dilemma (pp. 125–126)

1.) What is the current population trend in Russia today? What factors have accounted for this pattern?

2.) What are the current attitudes in Russia towards immigration from other countries?

Review Section 2.2.9   Russia's Prospects (pp. 126–128)

1.) Can be described as either a 'heartland' or a 'rimland', or both? Why?

2.) What are the two main strategic concerns of Russia today?

3.) Why is it difficult to get a private company started in Russia today? What impact has this had upon exports from the Russian Realm?

4.) What two mineral resources are abundant in Russia, and what geopolitical issues are these causing?

5.) The most efficient oil pipeline for exporting petroleum in Southwest Russia would be located inside the borders of what present-day countries in the realm?

Review Section 2.2.10   Russia's Core and Peripheries (pp. 128–136)

1.) Describe the Core Area in terms of population, agriculture and economic impact on the realm.

2.) What are the two great cities found within this region today? Briefly describe each in terms of population and cultural/political influence, both past and present.

3.) Name two major port cities facing the Arctic Ocean in this region:

4.) What spatial impacts do the Ural Mountains have upon the region, and Russia as a realm?

Review Section 2.2.11   The Eastern Frontier (pp. 137–138)

1.) Describe the location of the Eastern Frontier region, and some of its associated physical features.

2.) Where is the 'Kuzbas' area, and what importance does it have in terms of mineral supplies?

3.) Describe the physical and population geography of this region east of Lake Baikal:

Review Section 2.2.12   Siberia (pp. 137–138)

1.) What main features characterize Siberia as a region, especially environmentally?

2.) Describe the human geography of Siberia, and how it differs from the rest of the realm.

Review Section 2.2.13   The Russian Far East (pp. 138–141)

1.) Even with all of its natural resources and strategic location, why has the eastern coastline of Russia not seen any kind of significant development?

2.) Name the two parts of this region, and describe the differences between them in terms of human settlement and physical features.

3.) What reasons have led to the post-Soviet transition being exceptionally difficult in this region?

Review Section 2.2.14   Transcaucasia: The External Southern Boundary (pp. 141–144)

1.) Name the three countries of Transcaucasia, and briefly describe their impact upon the realm economically and culturally:

A.)

B.)

C.)

SECTION 2.3:   Blank Maps of Russia

Blank maps are provided for you to utilize with the mapping questions from Section 2.1 of the Study Guide. You are being provided with one blank map for each mapping exercise given in Section 2.1. It is a good idea to make additional copies of these blank maps to use for extra practice and review.

SECTION 2.4:   Student Companion Website

Additional study tools are available on the Student Companion Website at www.wiley.com/college/deblij. Features include:

- *Flashcards* offer an excellent way to practice key concepts, ideas, and terms from the text. You can review and quiz yourself on the concepts, ideas, and terms discussed in each chapter.
- *Map Quizzes* help you to master the place names for the various regions studies. Three game-formatted map activities are provided for each chapter.
- *Chapter Review Quizzes* provide immediate feedback to true/false, multiple choice, and short answer questions.
- *Audio Pronunciation* is provided for over 2000 key words and place names from the text.
- *Annotated Web Links*
- *Area and Demographic Data*

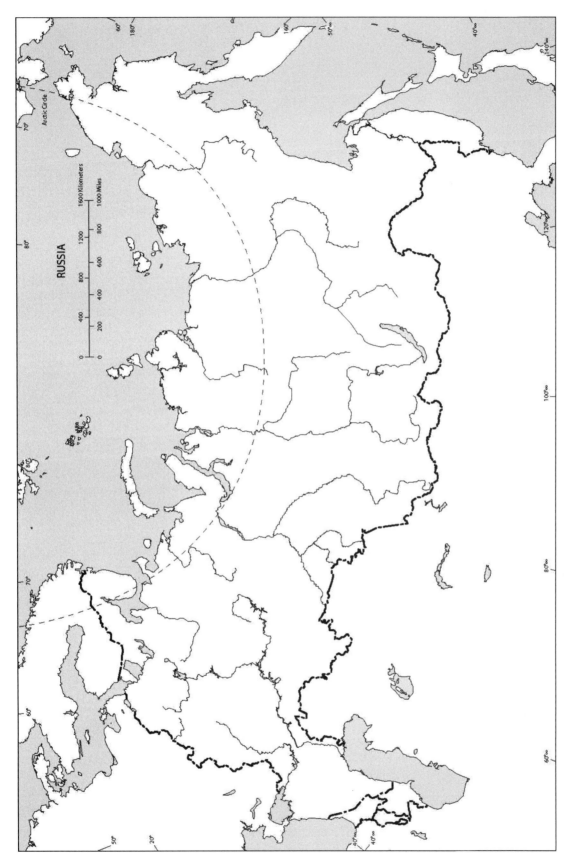

RUSSIA

Arctic Circle

1600 Kilometers
1000 Miles

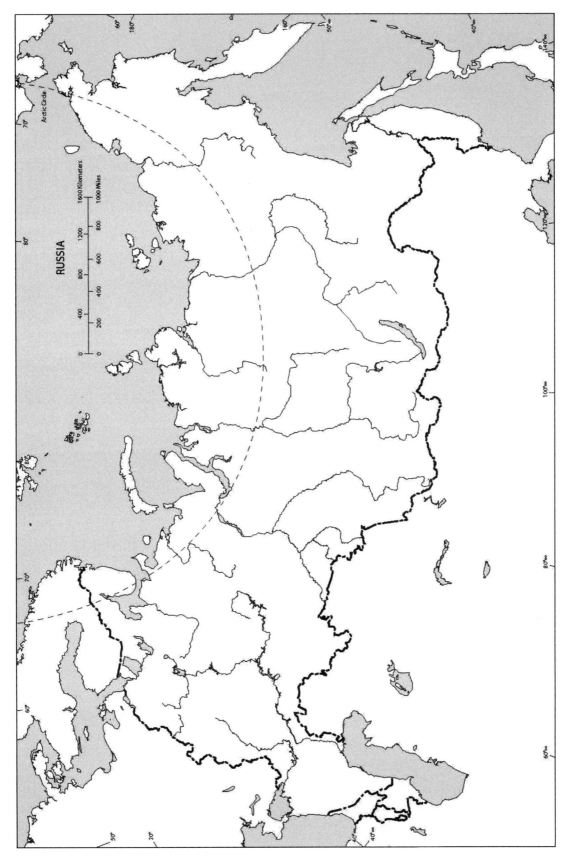

RUSSIA

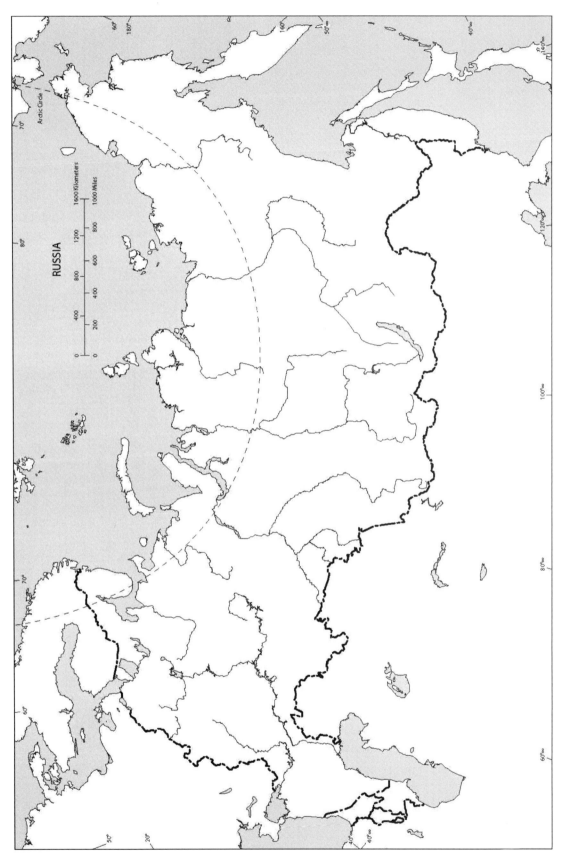

RUSSIA

Arctic Circle

1600 Kilometers
1000 Miles

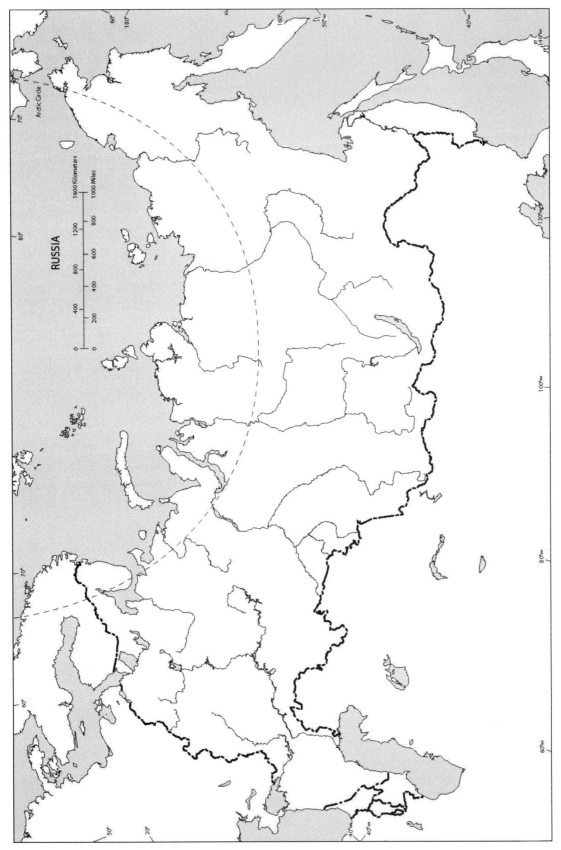

RUSSIA

Arctic Circle

1600 Kilometers
1000 Miles
1200
800
800
600
600
400
400
200
200
0
0

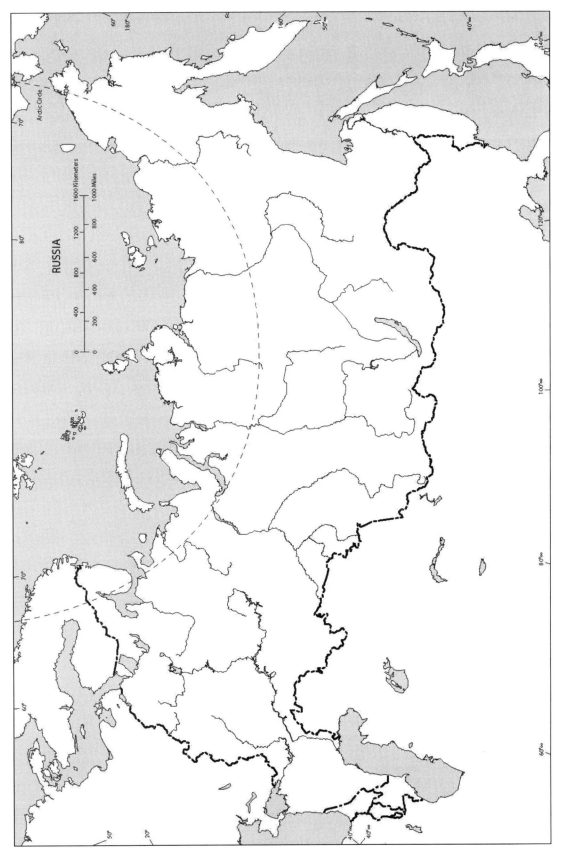

RUSSIA

Arctic Circle

1600 Kilometers
1000 Miles

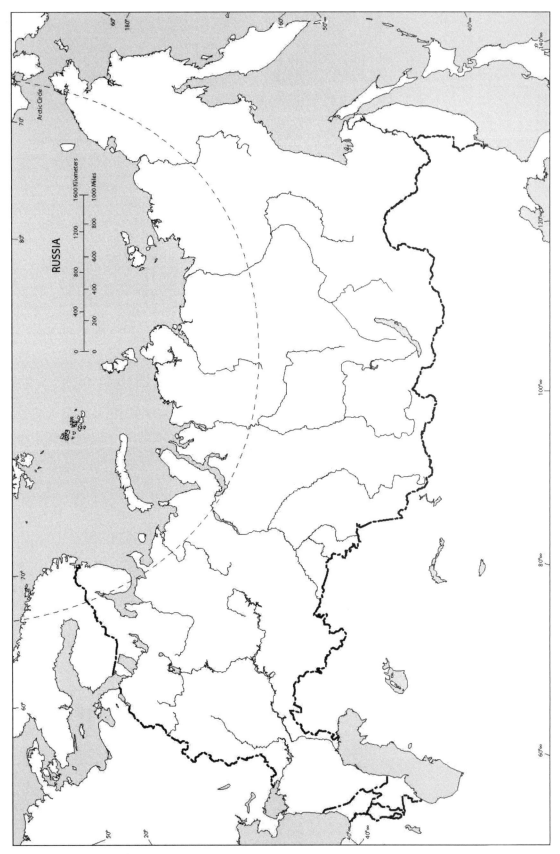

RUSSIA

1600 Kilometers
1000 Miles

Arctic Circle

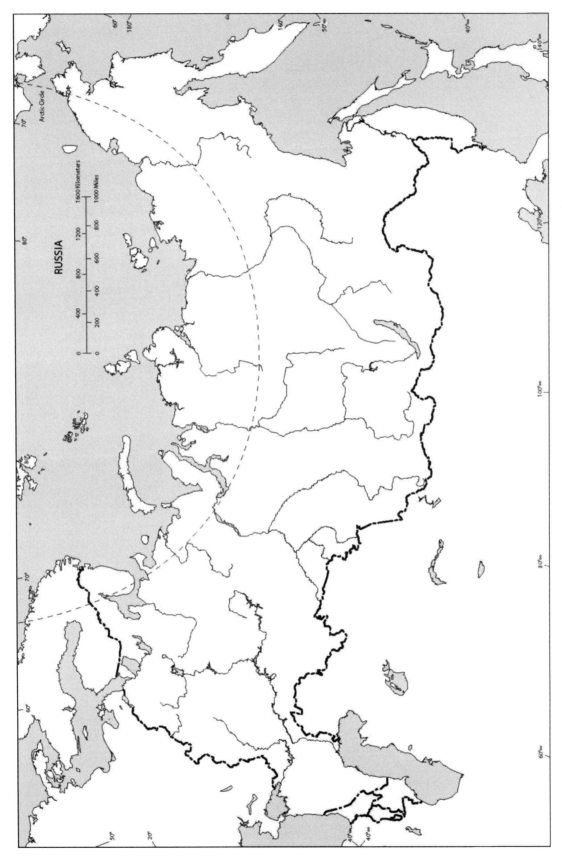

RUSSIA

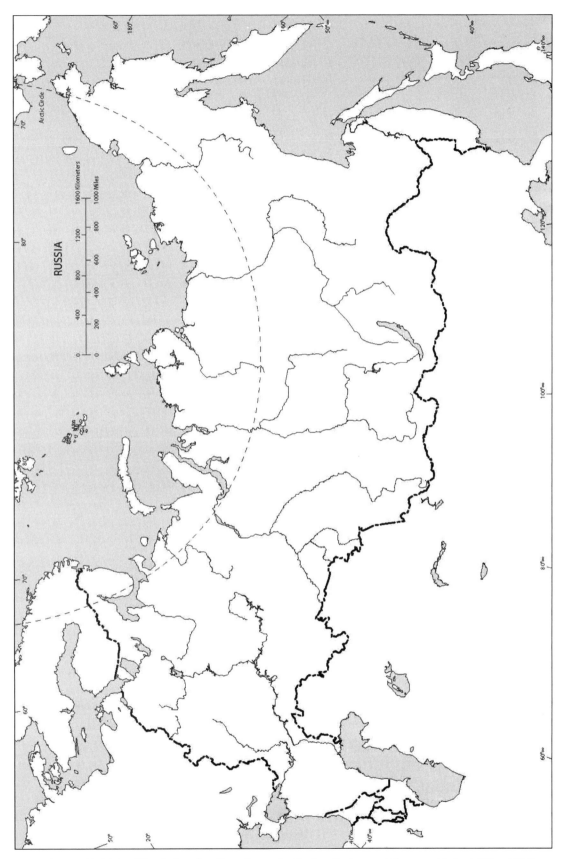

RUSSIA

Arctic Circle

1600 Kilometers
1000 Miles

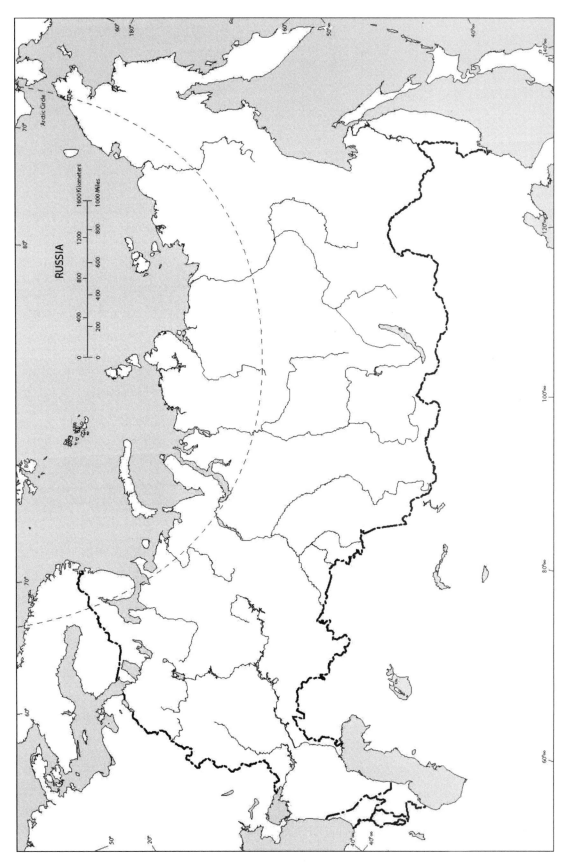

RUSSIA

CHAPTER #3   NORTH AMERICA

SECTION 3.1:   Map Creation

In this section you will create your own detailed maps of the different major physical, cultural, geopolitical and environmental attributes of the realm. Each of these maps is critically important for understanding the Major Geographic Qualities of North America, including their spatial distribution and functionality. By creating these on your own using the blank maps provided at the end of this chapter, you will actively learn where these different features are located and how their spatial distribution affects the ways of life in each of these respective areas.

Blank Maps for North America are provided for you in Section 3.3. Label each of your maps with the appropriate name and title. For instance, the first map you will create will be called 'Map 3–1: States of the USA and Provinces of Canada'.

Map 3–1:   States of the USA and Provinces of Canada

On the map, label the following States of the USA and Provinces of Canada. These are listed in alphabetical order, by country. You may need to use abbreviations and/or arrows if you do not have enough space to write in the full name on the map. To assist with creation, refer to Figure 3–6, p. 158.

States of the USA:

- Alabama	- Hawaii	- Massachusetts	- New Mexico	- South Dakota
- Alaska	- Idaho	- Michigan	- New York	- Tennessee
- Arizona	- Illinois	- Minnesota	- North Carolina	- Texas
- Arkansas	- Indiana	- Mississippi	- North Dakota	- Utah
- California	- Iowa	- Missouri	- Ohio	- Vermont
- Colorado	- Kansas	- Montana	- Oklahoma	- Virginia
- Connecticut	- Kentucky	- Nebraska	- Oregon	- Washington
- Delaware	- Louisiana	- Nevada	- Pennsylvania	- West Virginia
- Florida	- Maine	- New Hampshire	- Rhode Island	- Wisconsin
- Georgia	- Maryland	- New Jersey	- South Carolina	- Wyoming

Provinces of Canada:

- Alberta	- Nunavut
- British Columbia	- Ontario
- Manitoba	- Prince Edward Island
- New Brunswick	- Quebec
- Newfoundland and Labrador	- Saskatchewan
- Nova Scotia	- Yukon
- Northwest Territories	

Map 3–2: Major Cities of North America

On this map, label the following major cities of the North American Realm. You may need to use abbreviations and/or arrows if you do not have enough space to write in the full name on the map. To assist with creation of this map, refer to the overall map of North America found on p. 146 and also Figure 3–6, p. 158.

Major Eastern USA Cities:

- Boston	- New York City	- Philadelphia	- Hartford
- Atlantic City	- Manchester (NH)	- Burlington	- Portland (ME)
- Trenton	- Washington D.C.	- Baltimore	- Wilmington
- Annapolis	- Pittsburgh	- Providence	- Richmond
- Charleston SC	- Charleston WV	- Buffalo	- Syracuse
- Albany	- Charlotte	- Raleigh	- Norfolk

Major Northern and Midwestern USA Cities:

- Cleveland	- Cincinnati	- Louisville	- Indianapolis
- Chicago	- Detroit	- Lansing	- Lexington
- Milwaukee	- Madison	- Minneapolis	- Des Moines
- Springfield	- St. Louis	- Green Bay	- Duluth
- Bismarck	- Pierre	- Kansas City	- Omaha
- Billings	- Helena	- Fargo	- Wichita

Major Southern USA Cities:

- Miami	- Tampa	- Orlando	- Jacksonville (FL)
- Atlanta	- Savannah	- Chattanooga	- Nashville
- Columbia	- Birmingham	- Montgomery	- Jackson (MS)
- New Orleans	- Memphis	- Houston	- Dallas
- San Antonio	- El Paso	- Oklahoma City	- Tulsa
- Little Rock	- Knoxville	- Austin	- Amarillo

Major Western USA Cities:

- Albuquerque	- Santa Fe	- Denver	- Cheyenne
- Casper	- Colorado Springs	- Phoenix	- Tucson
- Las Vegas	- Salt Lake City	- Reno	- Boise
- San Diego	- Los Angeles	- San Francisco	- Sacramento
- Portland (OR)	- Seattle	- Spokane	

Major Canadian Cities:

- Vancouver	- Calgary	- Edmonton	- Victoria
- Regina	- Winnipeg	- Thunder Bay	- Toronto
- Windsor	- Ottawa	- Montreal	- Quebec City
- Charlottetown	- Fredericton	- Halifax	

Map 3–3: Physical Features of North America

North America has a wide diversity of physical features throughout the realm that have played an important role in settlement patterns, political boundaries, distribution of mineral resources, and environmental issues as well. For this map, you will draw in the physical features listed below. These have been divided for you into categories, and it is recommended to use a different colored pencil or pen for each of the names per category. To assist with creation of this map, refer to both the overall map of North America found on p, 146 and also Figure 3–3, p. 152.

Mountain Ranges: Using a blue pen/pencil, draw in the locations of these mountain ranges on your map and label them accordingly. Use a Λ symbol to denote their locations.

- Appalachian Mountains
- Adirondack Mountains
- Cascade Mountains
- Green Mountains
- Ozark Mountains
- Wasatch Mountains
- White Mountains
- Rocky Mountains

Physiographic Regions: These regions all have a physical characteristic in common, whether being lowlands or islands. Using Figure 3–3 (p. 152) as guide, circle and label the following physical regions on your map using a green pen/pencil:

- Gulf – Atlantic Coastal Plain
- Interior Lowlands
- Interior Highlands
- Intermontane Basins and Plateaus
- Piedmont
- Canadian Shield
- Rocky Mountains
- Pacific Mountains and Valleys
- Appalachian Highlands
- Arctic Coastal Plain
- Great Plains

Map 3–4: North American Bodies of Water

North America has been influenced by its water resources, both in terms of spatial history and also economic development. On this map, you will identify the major bodies of water found inside this realm. These are divided into groups and it is recommended to use a different colored pencil or pen for each of the names per water body category. Use p. 146 as a guide.

Major Saltwater Features:
- Atlantic Ocean
- Pacific Ocean
- Gulf of Mexico
- Arctic Ocean

Major Water Bodies in the Eastern United States:
- Mississippi River
- Lake Ontario
- Lake Superior
- Ohio River
- Lake Erie
- Lake Okeechobee
- Missouri River
- Lake Michigan
- Tennessee River
- Hudson River
- Lake Huron
- Red River

Major Water Bodies in the Western United States:
- Rio Grande River
- Salton Sea
- Colorado River
- Sacramento River
- Great Salt Lake
- Puget Sound
- Columbia River
- San Francisco Bay

Major Water Bodies in Canada:
- St. Lawrence River
- Baffin Bay
- Vancouver Bay
- Lake Winnipeg
- Great Bear Lake
- Bay of Fundy
- Mackenzie River
- Great Slave Lake
- Hudson Bay
- Ottawa River

Map 3–5: *Major Climate Regions of North America*

North America is a realm that contains a wide variety of climates. Most of these climates contain very cold temperatures throughout the year, although there are some slightly warmer climates found in the extreme southern portions of the realm. Since there is not a formal climate map for this realm in the chapter, refer to the World Climate Map (Figure G – 7, pp. 16–17).

The major Köppen Climate types are given below for each of the climates found in this realm. Using a different colored pen/pencil for each category, outline and then color in these climate types:

Highland Climates: *Tropical Climates*:
- H – Climates   - Aw

Mild Mid-Latitude Climates: *Cold Mid-Latitude Climates:*
- Cfa   - Dfa
- Csa   - Dfb
- Cfb   - Dfc

Dry Climates:
- BWh   - BWk
- BSh   - BSk

Map 3–6: *North American Population Densities*

Examining the North American Population Distribution map on Figure 3–2 (p. 150), we see that four population density categories exist in the Realm. Using a different colored pen/pencil for each category, outline and color the countries/sections of countries that fit into each of these designations. Once completed, you will see a spatial pattern of population distribution emerge across the realm.

Very Dense: This category is found in two locations:

• Along the Northeast USA coastline from Washington D.C. to Boston

• In the Southwestern USA, from San Diego through Los Angeles

Dense: This population category accounts for the rest of Eastern North America from the Atlantic Coast to the Mississippi River. It also includes Eastern Texas, the rest of California, and the Northwest from Portland (OR) to Vancouver (B.C.).

Sparse: Found in several locations: Great Plains of the USA and Canada, The Desert Southwest including Eastern Oregon and Idaho, and the Northeast USA and Canada, from Vermont to Newfoundland.

Very Sparse: This includes the states of Montana and the Dakotas, along with everything in Canada that is about 100 miles north of the USA border.

Map 3–7:   Spatial History: The Evolution of the North American Union

Using Figure 3–5 (p. 156) and Figure 3–13 (p. 176), you will create a map that shows the spatial development of the North American Union. As time progressed from the first colonies until present-day, North America developed spatially by starting in the east and progressing westward. However, the rate of development was much faster in the USA than Canada, as evidenced by the dates each state or province was respectively admitted as part of the union.

Use your map to draw in the boundaries of settlement by their respective dates in the North American Realm, both for Canada and the United States of America. Categories for dates admitted to either the USA or Canada have been established below, for a more clear settlement picture.

Category A: *States of the USA, by 1800*

Category B: *States of the USA that joined between 1801–1830*

Category C: *States of the USA that joined between 1831–1860*

Category D: *States of the USA that joined between 1861–1899*

Category E: *States of the USA that joined from 1900 to present day*

Category F: *Provinces of Canada that joined the nation between 1867–1899*

Category G: *Provinces of Canada that joined the nation from 1900 – Present Day*

Map 3–8:   North American Religious Distribution

Using Figure 3–11 (p. 168), you will create a map that shows the spatial distribution of religions concentrations in North America. Though primarily Christian, North America contains a wide variety of sects of Christianity as well as a minor presence of tribal religions in the realm.

Catholic: Found in the Northeast USA and Quebec; also found in the Southwest USA from New Mexico to Arizona and Southern California.

Baptist: In the Southeast USA primarily, from Virginia through the 'Cotton Belt' to Texas.

Methodist: Found primarily in the Midwest, including the states of Ohio, Indiana, Illinois and Iowa.

Amish: Two primary sects (277,000 practicing Amish today): SE Pennsylvania (centered around the town of Lancaster), and NE Indiana/NW Ohio, from Elkhart to Toledo.

Lutheran: Found in Northern Wisconsin, Minnesota, the Dakotas, Manitoba and Saskatchewan.

Mormon: Centered in Utah primarily, though has diffused some to Western Nevada.

Tribal: Found in Northern Canada and in the Desert Southwest of the USA, associated with Native American populations.

Map 3–9: Agricultural Activity in North America

Using Figure 3–15 (p. 180) as a guide, circle and label the following agricultural regions of North America on your map. Additionally, write in the primary agricultural product of that region on your map. For instance, The Heartland's primary product would be grain. Use a different color pen/pencil (if possible) for each area, and consult the textbook for further information on these locations.

- The Basin and Range
- The Heartland
- The Northern Great Plains

- The Eastern Uplands
- Mississippi Portal
- Prairie Gateway

- The Fruitful Rim
- Northern Crescent
- Southern Seaboard

Map 3–10: Energy and Mineral Regions in North America

Using Figure 3–7 (p. 160) as a guide, circle and label the following areas on your map. Some of the mineral deposit locations tend to be scattered around the map, in which case do your best to enclose the primary general location(s) of activity. Use a different color pen/pencil (if possible) for each area, and consult the textbook for further information on these areas.

- Coal Mining
- Copper Mining

- Oil Mining
- Nickel Mining

- Natural Gas Mining
- Lead Mining

- Gold Mining
- Silver Mining

Map 3–11: Major Urban Regions in North America

Using Figure 3–9 (p. 163) as a guide, circle and label the following economic/urban areas on your map. Use a different color pen/pencil (if possible) for each area, and consult the textbook for further information on these areas.

The North American Manufacturing Belt: This is/was the primary location of manufacturing in the USA from the early 1800's until the mid-1900's. Today this region is also known as the 'Rust Belt', due to all of the decommissioned factories rusting shut.

Primary Urban Growth Areas: These constitute the 'major' cities of North America, both in historic and present times.

Secondary Urban Growth Areas: These are regions of newer urban growth and expansion, and tend to be more centrally located in North America, compared to the primary growth regions which are more oriented along the coastlines of the realm.

*Map 3–12:   **Environmental Issues in North America***

The North American Realm, while one of the most advanced today in terms of environmental protection and regulations, still contains some serious issues that threaten its land. These environmental issues are divided into categories below. Using a different colored pen/pencil for each, circle or highlight the areas and label them using their respective environmental situation.

Groundwater Depletion: The Great Plains region (Texas to Nebraska) has seen severe problems with over-pumping groundwater since the late 1800's, primarily due to irrigation for agriculture. The is named the Ogallala Aquifer, and is one of the most threatened in North America today.

Coastal Pollution: Mostly found along the eastern coastline of the USA, from Miami to Boston. Coastal Pollution is also a problem in the Gulf of Mexico from Louisiana to Texas (oil drilling), and from San Diego to San Francisco, CA.

River Pollution: Most of North America's major rivers have been cleaned up substantially since environmental protection laws commenced in the 1970's. However, river pollution is still a concern on the waterways with the highest density of commercial traffic. These rivers include the Mississippi, Missouri, Hudson, Ohio, Rio Grande, Columbia, and Colorado Rivers.

Acid Rain: Manufacturing and the resulting air pollution has been responsible for tremendous problems with air quality and acid rain in Eastern North America. Although environmental regulations have curbed this problem significantly, and though manufacturing has been declining in the realm in the last 30 years, the Northeast USA and Eastern Canada are the areas still most affected by acid rain today.

Urban Air Quality: The highest densities of urban air pollution are found in the following locations:

- The Northeast USA (Washington D.C. – Boston)
- Southeast USA (Charlotte NC – Atlanta GA – Birmingham AL)
- Southwest USA (San Diego – Los Angeles)
- Texas 'Triangle' (Dallas – San Antonio – Houston)

Map 3–13:   *Ethnic Groups in North America*

In this map, you will show the spatial distribution of ethnic groups found throughout North America. With Figure 3–14 (p. 177) and the descriptions below serving as a guide, circle and label the following areas on your map and use a different color pen/pencil (if possible) for each cultural region.

African – Americans:	Primarily in the Southeast USA, from Virginia through North and South Carolina, Tennessee, Georgia, Alabama, Mississippi, Arkansas, Louisiana, and Eastern Texas.
Hispanics:	Two main locations: Florida (Hispanics of Puerto Rican and Cuban origin), and the Southwest USA from Texas to California (Hispanics of Mexican origin).
Asians:	The primary concentration is from San Diego north along the Pacific Coast to Vancouver (B.C.). Another concentration exists in the Toronto (Ontario) region.
French Canadians:	Located within the Province of Quebec in Canada.
Native Americans:	Found within the following locations in the realm:

- The Desert Southwest, centered on the 'Four Corners' region where the states of Arizona, New Mexico, Colorado and Utah meet.

- North and South Dakota, and Eastern Montana

- Eastern Oklahoma

- Northern Canada, primarily in the provinces of Yukon, The Northwest Territories, Nunavut, and then in Northern Quebec.

Map 3–14:   *Regions of the North American Realm*

Using Figure 3–16 (p. 182) in the textbook, color the following regions of the North American Realm and use a different colored pen/pencil for each region and the country list below as a guide. After completing this exercise, consult the textbook to understand why each of these is considered a separate region of the realm.

- The Core	- The Maritime Northeast	- French Canada
- The South	- The Southwest	- The Pacific Hinge
- The Western Frontier	- The Continental Interior	- The Northern Frontier

SECTION 3.2:   Review Questions

In this section you will review the main concepts and terminology from the North American Realm by answering the questions provided below. Write in your answers in the space provided, and refer to the respective sections/pages of the chapter for the correct information.

Review Section 3.2.1   North America's Major Geographic Qualities (p. 148)

Summarize the nine Major Geographic Qualities of North America in your own words:

1.)

2.)

3.)

4.)

5.)

6.)

7.)

8.)

9.)

Review Section 3.2.2   Defining the Realm (pp. 148–149)

1.) In terms of Human Geography, why is the continent of North America divided into two realms?

2.) What effects did 'American Indians' had on the realm, before European settlement?

3.) Describe the relationship between the USA and Canada, in terms of trade across the border:

Review Section 3.2.3   Population Growth and Clustering (pp. 150–151)

1.) What are the differences between the USA and Canada in terms of their respective population distributions?

2.) What is *Cultural Pluralism*?

3.) What is the current population of the USA? What is its projected population in 2040? What factors would lead to this increase in population?

Review Section 3.2.4   North America's Physical Geography (pp. 151–154)

1.) Name at least five of the major physiographic provinces of North America. Where are they found, and what are their major attributes?

   A.)

   B.)

   C.)

   D.)

   E.)

2.) Briefly describe the variation in North America's climates:

3.) How do climates change as you move away from the ocean into the interior of the continent?

4.) What is the *Rain Shadow Effect*?

5.) What are the two major drainage systems in this realm, and where are they located?

6.) What are the *Fall Line Cities* and where are they located? Give at least three examples:

Review Section 3.2.5   Indigenous North America (pp. 154)

1.) Who preceded the Europeans in arriving on the North American continent, how long ago did they arrive and where did they settle?

2.) Describe the impact that the European settlers had upon the Native American populations.

3.) About what percentage of USA Territory is set aside for Native American reservations today?

Review Section 3.2.6   The Course of European Settlement and Expansion (pp. 154–157)

1.) What were the motivations of the British for settling North America? How does this differ from the French motivation?

2.) What were the specialties of the three primary British colonies, in terms of products?

3.) What prompted people to migrate to the Interior Lowlands? What were some of the results of this migration?

4.) Describe the French colonization influence, in terms of where they settled both in present-day Canada and the USA:

5.) What moment started the beginning of the slave trade in North America?

6.) Describe the differences in settlement between the USA and Canada in the year 1850:

Review Section 3.2.7   The Federal Map of North America (pp. 157–159)

1.) How are the political geographies of Canada and the USA similar?

2.) In what ways are Canada and the USA different politically?

3.) From Figure 3.6, what are the largest cities in Canada in terms of population?

Review Section 3.2.8   North America's Natural Resources (pp. 159–160)

1.) What is an aquifer, and why would it be considered a natural resource?

2.) What is a fossil fuel? Where are the primary locations of fossil fuel deposits found in this realm?

Review Section 3.2.9   Population and Multiculturalism (p. 161)

1.) Describe the 'center of gravity' of population in the USA, in terms of the direction it has been moving since the 1800's. Why would it move that way?

2.) What made the 'Sunbelt' states attractive to people for relocation, especially starting in the 1970's?

3.) List the primary five reasons for continued migrating in the USA and Canada:

    A.)

    B.)

    C.)

    D.)

    E.)

4.) How is Canada's view on migration different than that of the USA? What ethnic groups have migrated there primarily since the 1990's?

Review Section 3.2.10   Cities and Industries (pp. 161–166)

1.) From p. 162, what are the three reasons that migrations vary in size?

A.)

B.)

C.)

2.) What are push and pull factors? Give examples of these in North America:

3.) Where is North America's 'Manufacturing Belt'? What types of manufacturing are/were taking place within this region?

4.) What is the 'Megalopolis'? Where is it located? Name at least five major cities inside of the region.

5.) Where is Canada's 'Main Street' located? Name at least three major cities found within this area.

6.) From p. 164, describe New York City in terms of its place both in North America as a center of population, and on the global stage in terms of its financial and economic importance:

7.) Describe the structure of the typical city in North America (particularly the USA), incorporating the terms Central Business District (CBD), Ghetto, Outer Cities and Suburban Downtowns into your explanation.

Review Section 3.2.11   Cultural Geography of North America (pp. 166–168)

1.) What characteristics do the USA and Canada share in common, in terms of cultural similarities? Give at least four different examples.

2.) What differences do Canadian cities possess, as opposed to their American counterparts?

3.) How is Canada different than the USA in terms of language, both spatially and politically?

4.) What role has religion played in the settlement of the USA and Canada? Give examples of common sects of Christianity found today throughout the realm.

Review Section 3.2.12   The Changing Geography of Economic Activity (pp. 169–170)

1.) What is the 'Postindustrial' Era? How does it relate to North America in terms of its economy?

2.) Briefly describe the following four components of the spatial economy:

    A.) Primary Activity

    B.) Secondary Activity

    C.) Tertiary Activity

    D.) Quaternary Activity

3.) Where is 'Silicon Valley' located? What types of goods and services does this region produce?

Review Section 3.2.13   Canada (pp. 170–175)

1.) How is Canada divided in terms of provinces and territories? What percentage of Canadians live in the provinces?

2.) Who are the Inuit People, and where are they located?

3.) Give examples of the mineral and energy resources found in Canada, and the locations in which these are primarily located.

4.) What is Canada's National Capital? Where is it located, both in terms of provinces and the river that separates these?

5.) What percentage of Canadians speak English? French? Other languages?

6.) Because of the language differences, Canada is divided along what lines?

7.) Why did the British Parliament divide the country into Upper and Lower Canada? Where are these located relative to one another?

8.) Name the three main reasons that support for Quebec's independence has weakened in recent years:

    A.)

    B.)

    C.)

9.) What is the name of the indigenous people in Northern Quebec? What percentage of Quebec's territory do they occupy, relative to the French-speaking people?

10.) What are some of Canada's economic strengths, and where are they found?

11.) What is NAFTA, and what effect has it had upon trade in North America, specifically Canada?

Review Section 3.2.14   The United States of America (pp. 175–181)

1.) What were the final two states to be added to the USA? When did they join the union?

2.) What influence did the Treaty of Guadalupe Hidalgo have upon the land area of the United States, relative to Mexico?

3.) What is the current population of the USA?

4.) Name some ethnic groups found in the USA, and the regions where you will find their respective greatest concentrations:

5.) From the Regional Issue discussion on Immigrants in North America (p. 178), briefly describe the pros and cons to immigration as presented in these opposing views. What is your opinion?

6.) What is a 'mosaic culture', and how does this apply to the USA?

7.) What percentage of workers in the USA are still employed by agricultural (i.e. farm) positions?

8.) Name at least five major crops in the USA and describe where they are located:

9.) What is an 'aerotropolis', and how could this term apply to cities in the USA?

Review Section 3.2.15   The Regions of North America (pp. 181–190)

Summarize each of the following Regions of North America below, listing their major spatial qualities in terms of location, physical and cultural features (if applicable), and economy:

1.) North American Core:

2.) The Maritime Northeast:

3.) French Canada:

4.) The South:

5.) The Southwest:

6.) The Pacific Hinge:

7.) The Western Frontier:

8.) Continental Interior:

9.) The Northern Frontier:

SECTION 3.3: Blank Maps of North America

Blank maps are provided for you to utilize with the mapping questions from Section 3.1 of the Study Guide. You are being provided with one blank map for each mapping exercise given in Section 3.1. It is a good idea to make additional copies of these blank maps to use for extra practice and review.

SECTION 3.4: Student Companion Website

Additional study tools are available on the Student Companion Website at www.wiley.com/college/deblij. Features include:

- *Flashcards* offer an excellent way to practice key concepts, ideas, and terms from the text. You can review and quiz yourself on the concepts, ideas, and terms discussed in each chapter.

- *Map Quizzes* help you to master the place names for the various regions studies. Three game-formatted map activities are provided for each chapter.

- *Chapter Review Quizzes* provide immediate feedback to true/false, multiple choice, and short answer questions.

- *Audio Pronunciation* is provided for over 2000 key words and place names from the text.

- *Annotated Web Links*

- *Area and Demographic Data*

NORTH AMERICA

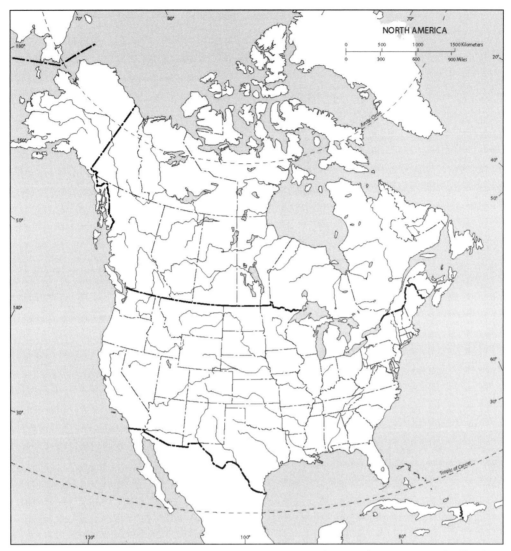

NORTH AMERICA

0   500   1000   1500 Kilometers
0   300   600   900 Miles

Arctic Circle

Tropic of Cancer

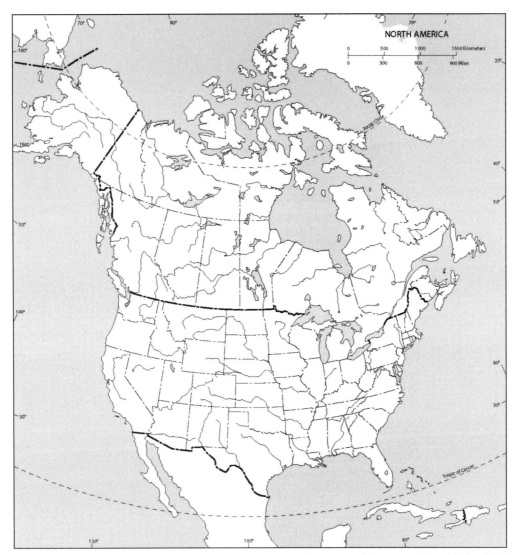

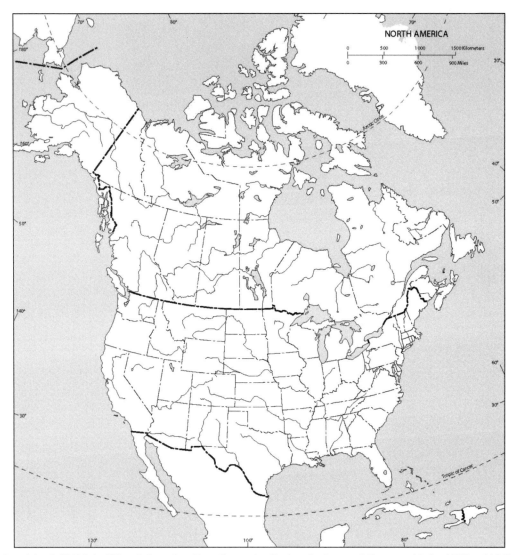

NORTH AMERICA

NORTH AMERICA

500 1000 1500 Kilometers
300 600 900 Miles

Arctic Circle

Tropic of Cancer

NORTH AMERICA

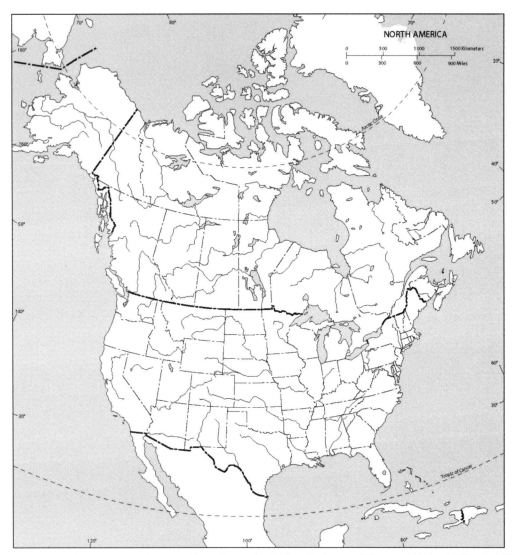

NORTH AMERICA

NORTH AMERICA

NORTH AMERICA

NORTH AMERICA

NORTH AMERICA

500 1000 1500 Kilometers
300 600 900 Miles

CHAPTER #4   CENTRAL AMERICA

SECTION 4.1:   Map Creation

In this section you will create your own detailed maps of the different major physical, cultural, geopolitical and environmental attributes of the realm. Each of these maps is critically important for understanding the Major Geographic Qualities of Central America, including their spatial distribution and functionality. By creating these on your own using the blank maps provided at the end of this chapter, you will actively learn where these different features are located and how their spatial distribution affects the ways of life in each of these respective areas. Blank Maps for Central America are provided for you in Section 4.3. Label each of your maps with the appropriate name and title. For instance, the first map you will create will be called 'Map 4–1: Countries of Central America'.

Map 4–1:   Countries of Central America

Label the following countries of the Central American Realm on your map. These have been divided into three regions, and you may need to use abbreviations and/or arrows if you do not have enough space to write in the full name on the map. To assist with creation of this map, refer to pp. 192–193.

Mainland Central America Countries:
- Mexico   - Guatemala   - Belize   - El Salvador
- Honduras   - Nicaragua   - Costa Rica   - Panama

Central American Countries in the Western Caribbean:
- Cuba   - Jamaica   - The Bahamas
- Puerto Rico   - Haiti   - The Dominican Republic

Central American Countries in the Eastern Caribbean:
- Aruba   - U.S. Virgin Islands   - British Virgin Islands
- Antigua & Barbuda   - Guadeloupe   - Dominica
- Martinique   - St. Lucia   - Barbados
- Grenada   - Trinidad and Tobago

Map 4–2:   Major Cities of Central America

On this map, label the following major cities of the Central American Realm. You may need to use abbreviations and/or arrows if you do not have enough space to write in the full name on the map. To assist with creation of this map, refer to the overall map of Central America found on pp. 192–193. Write in the city name using a different colored pencil or pen for each of the regions these belong.

Major Cities in Mainland Central America:
- Mexico City   - Tijuana   - La Paz   - Chihuahua
- Ciudad Juarez - Monterrey   - Guadalajara   - Acapulco
- Cancun   - Oaxaca   - Guatemala City   - San Salvador
- Tegucigalpa   - Managua   - San Jose   - Panama City

Major Cities in the Islands of Central America:
- Kingston   - Havana   - Guantanamo   - Nassau
- Port-Au-Prince   - Santo Domingo   - San Juan   - Port-of-Spain

Map 4–3:   Physical Features of Central America

For this map, you will draw in the physical features of Central America listed below. To assist with creation of this map, refer to the overall map of Central America found on pp. 192–193.

Mountain Ranges: Using a blue pen/pencil, draw in the locations of these mountain ranges on your map and label them accordingly. Use a Λ symbol to denote the locations of these features.
- Sierra Madre Occidental   - Sierra Madre Oriental   - Sierra Madre Del Sur
- Volcanic Axis of Central America (From Southern Mexico to Panama)

General Physical Features: These all are notable physical landscapes or features in the realm, whether peninsulas or groups of islands.
- Baja California Peninsula   - Yucatan Peninsula   - Nicoya Peninsula
- Panama Canal   - Azuero Peninsula   - Mosquito Coast
- Greater Antilles (i.e, the large islands of the Caribbean)
- Lesser Antilles (i.e. the small islands of the Caribbean)

Map 4–4:   Central American Bodies of Water

Central America is a realm that has been influenced by its water resources, both in terms of spatial history and also economic development. Unlike every other realm however, there is a noticeable lack of major rivers in the realm. On this map, you will identify the major bodies of water of Central America, Using pp. 192–193 as a guide.

Major Water Bodies on the Western Side of Central America:
- Pacific Ocean   - Gulf of California   - Gulf of Tehuantepec
- Lake Managua   - Lake Nicaragua   - Gulf of Foresca
- Gulf of Chiriqui   - Gulf of Panama

Major Water Bodies in the Caribbean:
- Caribbean Sea   - Gulf of Mexico   - Atlantic Ocean
- Yucatan Channel   - Strait of Florida   - Windward Passage
- Mona Passage   - Panama Canal   - Gulf of Mosquitos

Map 4–5:   Major Climate Regions of Central America

Central America is a realm that is mostly homogeneous in terms of climate. Since there is not a formal climate map for this realm in Chapter 4, refer to the World Climate Map (Figure G – 8) in the Introductory Chapter. The major Köppen Climate types are given below for each of the climates found in this realm. Using a different colored pen/pencil for each category, outline and then color in these climate types:

Tropical Climates:	*Dry Climates*:	*High Climates*:
- Af	- BWh	- H
- Aw	- BSh	

Map 4–6:   *Central American Population Densities*

Examining the Central American Population Distribution map on Figure 4–3 (p. 198), we see that three general population density categories exist in the Central American Realm. Using a different colored pen/pencil for each category, outline and color the countries/sections of countries that fit into each of these designations. This realm is unique in that it is the only realm of the world that does not contain very sparse population concentrations.

Very Dense: The highest population levels are found in three distinct areas:
- The exploding growth regions of Central Mexico and the Western Coastline from Southern Mexico to Nicaragua.
- The Greater Antilles, from Cuba to Puerto Rico.
- The Lesser Antilles

Dense: The rest of Mainland Central America will have the dense population category.

Sparse: The only sparse populations are found in Northern Mexico, due to a harsh climate and poor soils for agriculture.

Map 4–7:   *Spatial History: The European Influence on Central America*

Using Figure 4–6 (p. 202), you will create a map that shows the spatial impact of European Colonization in the Central American Realm. These countries originally settled the Caribbean for military and agricultural reasons, and recently these areas have become major tourist destinations. Use a different colored pen/pencil and show the following countries who colonized this area:

Britain: Settled the Mainland of Central America along the eastern shores from southern Mexico to Costa Rica. Also colonized The Bahamas, The Cayman Islands, Jamaica, and the southern Lesser Antilles.

French: Colonized Haiti and the northern half of the Lesser Antilles.

Spanish: Colonized the rest of Mainland Central America, along with Cuba, The Dominican Republic, and Puerto Rico.

Dutch: Looking for a region similar to the lucrative Spice Islands in Southeast Asia, the Dutch settled the small island regions near Aruba and also just east of Puerto Rico.

Map 4–8:   *The Cultures and People of Central America*

Using Figure 4–7 (p. 203), you will create a map that shows the spatial distribution of ethnic populations in the Central American Realm. This distribution is the result of centuries of European influence, both in driving the native people to other regions and also importing Africans as slaves to the realm to assist with agricultural ventures. Use a different colored pen/pencil and identify where these groups are located:

Amerindian Influence: Located on the Mainland of Central America, from southern Mexico into Honduras.

Mestizo Influence: Found throughout the rest of Mainland Central America, the exception being Northern Mexico and the eastern shorelines from Belize to Panama.

Spanish Influence:	Two distinct locations are seen in this realm: • Northern Mexico and Panama (Mainland) • Cuba and Puerto Rico

African Influence:	Primarily found along the eastern shoreline of the Mainland, from Belize to Panama.
European Influence:	Consisting of British, French and/or Dutch influences, these ethnic locations are found in Haiti, Jamaica and Trinidad and Tobago.

Map 4–9: Geopolitical Issues in Central America

In this map you will show where past and present political conflict areas are located. Using the descriptions below as your guide, identify the locations within Central America where you will find these situations.

Cuba:	Communist government, nearly the stage for nuclear war against the USA in 1962.
Haiti:	Significant numbers of refugees have fled the country to the USA in the last two decades; also a severe problem with disease as some locations in Haiti have 50% of their population with HIV/AIDS.
Honduras:	Location where Hurricane Mitch (180mph winds) slammed into Central America, causing widespread destruction, loss of life, and created thousands of environmental refugees along with disease issues.
Puerto Rico:	The government wants to become the 51st State of the USA; many citizens of Puerto Rico are not supportive of this idea.
Nicaragua:	The USA helped overthrow the dictator Manuel Noriega in 1990.
Panama:	Ownership of the Panama Canal reverted to Panama from the USA in 1999. If the government became unstable, it could disrupt worldwide shipping and the global economy.

Map 4–10: Environmental Issues in Central America

The Central American Realm is experiencing moderate to severe environmental issues in certain locations, for reasons ranging from mineral extraction to tourism and heavy shipping densities. These environmental issues are divided into categories below. Using a different colored pen/pencil for each, circle or highlight the areas and label them using their respective environmental situation.

Coastal Pollution:	Almost all of the Caribbean Sea is affected by coastal pollution, sources primarily being heavy shipping routes and also dumping of waste by cruise ships directly into the waters. The Panama Canal and the water on either side of it is heavily polluted for the same reasons.
Groundwater Pollution:	Mexico is infamous for having poor groundwater, with people getting sick due to lack of water treatment and pollution from both industrial and agricultural runoff. Groundwater pollution also exists in the Lesser

Antilles, as aquifers there are being drained for tourism and increased populations faster than they can be replenished.

Urban Air Quality: The industrial regions from Monterrey to Guadalajara and Mexico City in particular have significant issues with urban air quality. The smog in Mexico City is particularly noteworthy, as the lands surrounding the city represent a large 'bowl' being surrounded by mountains; therefore, the air pollution has a difficult time escaping the region.

Geologic Hazards: The region from Central Mexico to Panama is very geologically active, with numerous earthquake zones and active volcanoes lining the western coast of Mainland Central America.

Map 4–11:   Regions of the Central American Realm

Using Figure 4–2 (pp. 196–197) in the textbook, color the following regions of the Central American Realm and use a different colored pen/pencil for each region and the country list below as a guide. After completing this exercise, consult the textbook to understand why each of these is considered a separate region of the realm.

- Mexico   - Central America   - Greater Antilles   - Lesser Antilles

SECTION 4.2:   Review Questions

In this section you will review the main concepts and terminology from the Central American Realm by answering the questions provided below. Write in your answers in the space provided, and refer to the respective sections/pages of the chapter for the correct information.

Review Section 4.2.1   Central America's Major Geographic Qualities (p. 195)

Summarize the seven Major Geographic Qualities of Central America in your own words:

1.)

2.)

3.)

4.)

5.)

6.)

7.)

Review Section 4.2.2   Defining the Realm (pp. 195–197)

1.) Briefly describe the physiographic and cultural diversity of Central America, as summarized in this section of the textbook.

2.) What impact do culturally-based names such as 'Latin America' have upon a region?

3.) Why would Middle America warrant distinction as its own realm, instead of being considered a part of either North or South America?

4.) Name the four regions for now, and briefly describe how they may be different from one another just based on location and size of countries alone.

Review Section 4.2.3   Physiography of Central America (pp. 197–198)

1.) Describe the 'land bridge' of Middle America and its location. Why is it important?

2.) How many islands are there (approximately) in the Caribbean Sea? What are the names of the four largest islands?

3.) Define the term 'Archipelago', and how it relates to this realm:

Review Section 4.2.4   The Legacy of Mesoamerica (pp. 199–200)

1.) What is a 'Cultural Hearth'? What improvements and/or advancements did the cultural hearth in this realm offer to human civilization?

2.) Where is 'Mesoamerica'? Describe it in terms of geographic location and historical development.

3.) Compare and contrast the Maya and Aztecs in terms of spatial location, population densities (historical and present), and contributions to the realm culturally and technologically.

Review Section 4.2.5   Collision of Cultures (pp. 200–201)

1.) Describe the effects of Spain's defeat of the indigenous peoples in Middle America, giving at least four examples of the disaster that ensued for this culture:

A.)

B.)

C.)

D.)

2.) What were the rural, urban and religious impacts of the Spaniard conquest of Middle America? Give examples from the textbook.

Review Section 4.2.6   *Mainland and Rimland (pp. 201–204)*

1.) In which countries was the Spanish presence felt the most in Middle America? Were there any other European countries that also settled the realm, and if so, where were they located?

2.) What are the differences between the 'Mainland' and the 'Rimland' of Middle America? Give examples of countries that belong to each group.

3.) Who are the 'Mestizo' people? Why did they gain a presence in this realm?

4.) Plantations and Haciendas were the most common forms of agricultural systems in the realm. Describe each in terms of what they are, where they are found, and the types of agriculture in each.

5.) What are the five main differences between Plantations and Haciendas?

   A.)

   B.)

   C.)

   D.)

   E.)

Review Section 4.2.7   *Political Fragmentation (pp. 204–205)*

1.) Describe the political and cultural fragmentation in present-day Middle America:

2.) When was independence gained by countries in this realm? Is there a difference in time of independence between the Mainland and Rimland countries?

Review Section 4.2.8   *Mexico (pp. 205–215)*

1.) What geographic reasons allow Mexico to be considered its own region? Give at least five examples.

2.) Describe the physiography of Mexico, and how it differs somewhat from that of the United States.

3.) Where are the majority of the populations located in this region? Are there any parts of Mexico that contain sparse populations?

4.) What effect have push and pull factors had upon the populations and migrations in Mexico? Where are people moving, and why?

5.) Why would Mexico be considered a region of 'transculturation', instead of one of 'acculturation'? Define each term and state the reasons for this difference in Mexico.

6.) Summarize Mexico City in terms of population, cultures, economic activity and environmental issues currently facing the metropolis.

7.) What effect did the 1910 Revolution have upon the geography of Mexico?

8.) Describe the effect of 'ejidos' in Mexico, and the role of the government in these land areas.

9.) Name the thirteen sub-regions of Mexico, and briefly describe where each is located:

10.) Why is Mexico considered in 'progressive transition in many spheres'? Give examples from the textbook to support your answer.

11.) What are the main industrial activities in Mexico today? How do 'Maquiladoras' assist the region's economy, and where are they located?

12.) Describe the pros and cons of NAFTA in Mexico, specifically referring to the effect upon the workers of this realm.

13.) What does Mexico's future appear to be like, both economically and politically? What potential problems in the future could affect this region?

Review Section 4.2.9   The Central American Republics (pp. 215–222)

1.) Name the countries of the Central American Republics. How do these differ physically from Mexico?

2.) Where do most of the populations reside within this region? Why?

3.) What economic and political issues is this region currently facing?

4.) Building on Question 1, name and now briefly describe each of the seven countries of this region in terms of physical characteristics, population and cultures, economically and politically. What issues presently face each of these countries?

A.)

B.)

C.)

D.)

E.)

F.)

G.)

5.) Summarize the problem of tropical deforestation in the region, including the causes and extent of the situation. What solutions can you offer to this issue, if any?

6.) Summarize the Panama Canal in terms of when it was built and by what countries, where it is located including length, and the political and environmental issues facing the canal today.

Review Section 4.2.10   Altitudinal Zonation (pp. 215–216)

1.) Altitudinal Zonation refers to the idea that climates and vegetation patterns change as you increase in elevation going up the side of a mountain. (In other words, it gets colder as you go higher up a mountain, and the associated vegetation changes as well).
Name the five levels of Altitudinal Zonation below, and describe each in terms of their elevation, climate type, and vegetation that is grown (if applicable):

A.)

B.)

C.)

D.)

E.)

Review Section 4.2.11   The Caribbean Basin (pp. 223–230)

1.) What are the differences between the Greater and Lesser Antilles? Describe where each is located, and give examples of countries found within each group.

2.) Describe the economic situation of The Caribbean Basin region. Has it become better or worse over time, and why? Give examples to support your answer.

3.) Who are the 'Mulatto' people, and where are they located?

4.) Why is this region culturally heterogeneous? Give examples of different countries in the world that have either settled this region or had people migrate here since colonial times.

5.) What impact can tourism have upon the region, and where would be the focal points? Why would this region be considered desirable for tourism, and from what countries would the tourists originate?

6.) List and describe the five countries of The Greater Antilles, briefly discussing their physical attributes, cultures, economy and current issues they may be facing (whether political and/or environmental):

A.)

B.)

C.)

D.)

E.)

7.) What are the 'two clusters' of The Lesser Antilles, and how are they different? Give examples of countries found within each cluster.

SECTION 4.3: Blank Maps of Central America

Blank maps are provided for you to utilize with the mapping questions from Section 4.1 of the Study Guide. You are being provided with one blank map for each mapping exercise given in Section 4.1. It is a good idea to make additional copies of these blank maps to use for extra practice and review.

SECTION 4.4: Student Companion Website

Additional study tools are available on the Student Companion Website at www.wiley.com/college/deblij. Features include:

- *Flashcards* offer an excellent way to practice key concepts, ideas, and terms from the text. You can review and quiz yourself on the concepts, ideas, and terms discussed in each chapter.

- *Map Quizzes* help you to master the place names for the various regions studies. Three game-formatted map activities are provided for each chapter.

- *Chapter Review Quizzes* provide immediate feedback to true/false, multiple choice, and short answer questions.

- *Audio Pronunciation* is provided for over 2000 key words and place names from the text.

- *Annotated Web Links*

- Area and Demographic Data

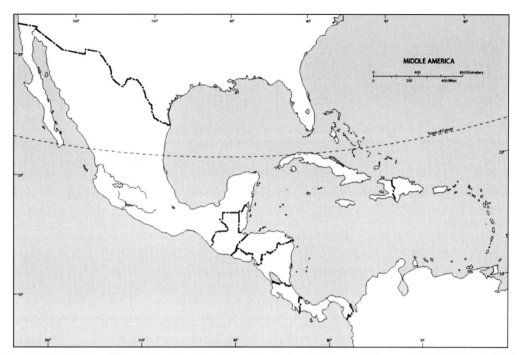

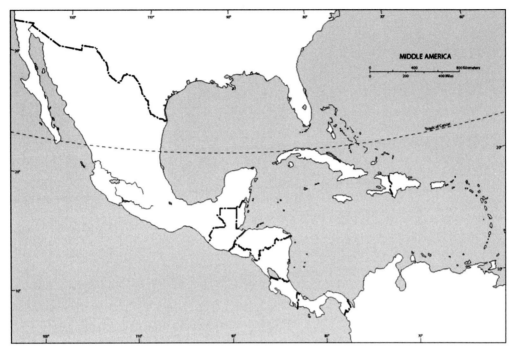

MIDDLE AMERICA

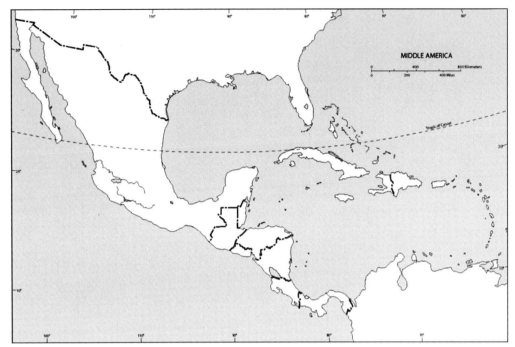

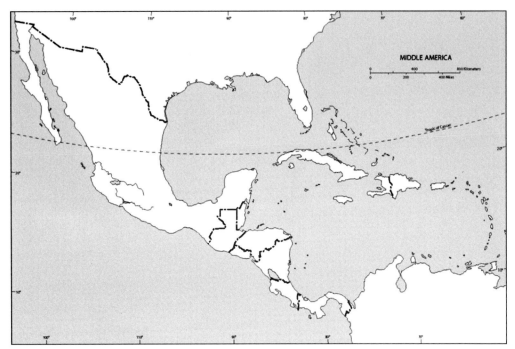

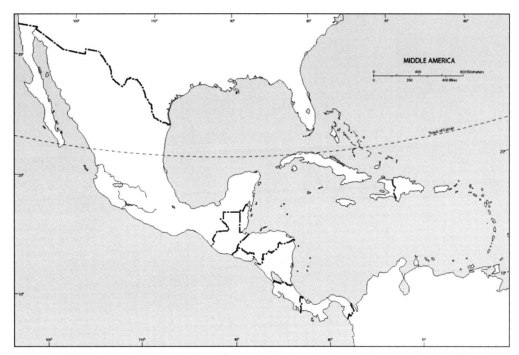

MIDDLE AMERICA

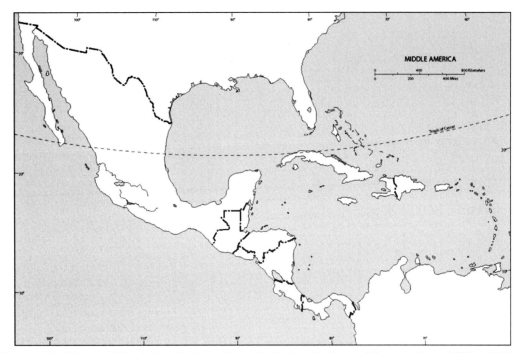

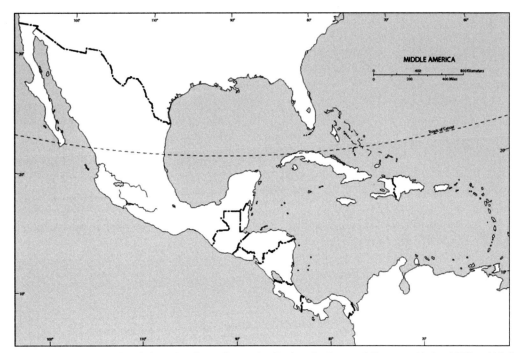

MIDDLE AMERICA

Tropic of Cancer

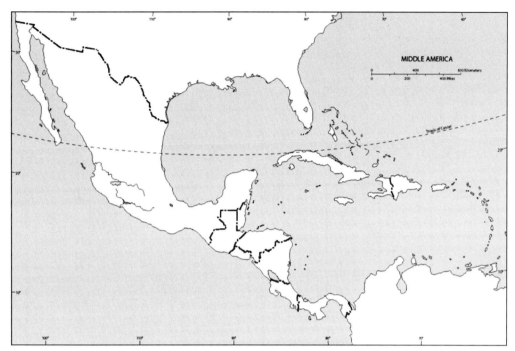

MIDDLE AMERICA

CHAPTER #5   SOUTH AMERICA

SECTION 5.1:   Map Creation

In this section you will create your own detailed maps of the different major physical, cultural, geopolitical and environmental attributes of the realm. Each of these maps is critically important for understanding the Major Geographic Qualities of South America, including their spatial distribution and functionality. By creating these on your own using the blank maps provided at the end of this chapter, you will actively learn where these different features are located and how their spatial distribution affects the ways of life in each of these respective areas. Blank Maps for South America are provided for you in Section 5.3. Label each of your maps with the appropriate name and title. For instance, the first map you will create will be called 'Map 5–1: Countries of South America'.

Map 5–1:   Countries of South America

On the map, label the following countries of the South American Realm. You may need to use abbreviations and/or arrows if you do not have enough space to write in the full name on the map. To assist with creation of this map, refer to p. 234.

Countries of South America:
- Brazil   - Paraguay   - Uruguay   - Argentina   - Chile
- Venezuela   - Guiana   - Bolivia   - Ecuador   - Peru
- Colombia   - French Guiana   - Suriname

Map 5–2:   Major Cities of South America

On this map, label the following major cities of the South American Realm. You may need to use abbreviations and/or arrows if you do not have enough space to write in the full name on the map. To assist with creation of this map, refer to the overall map of South America found on p. 234. Write in the city name using a different colored pencil or pen for each of the regions these belong.

Brazil Cities:
- Rio de Janeiro   - Brasilia   - Sao Paulo   - Recife
- Porto Alegre   - Manaus   - Porto Velho   - Fortaleza
- Salvador   - Macapa

Southern and Western Cities in South America:
- Montevideo   - Buenos Aires   - Asuncion   - Santa Cruz
- La Paz   - Tucuman   - Cordoba   - Rosario
- Bahia Blanca   - Santiago   - Antofagasta   - Lima

Northern Cities in South America:
- Quito   - Bogota   - Medellin   - Cartagena
- Caracas   - Maracaibo   - Georgetown   - Paramaribo

Map 5–3: Physical Features of South America

In this map, you will draw in the physical features of South America listed below. These have been divided for you into categories, and it is recommended to use a different colored pencil or pen for each of the names per category. To assist with creation of this map, refer to the overall map of South America found on p. 234.

Mountain Ranges: Using a blue pen/pencil, draw in the locations of these mountain ranges on your map and label them accordingly. Use a Λ symbol to denote the locations of these features.
- Andes Mountains   - Guiana Highlands   - Brazilian Highlands

General Physical Regions: These regions all have a physical characteristic in common, whether being plateaus, basins or islands. Using p. 234 as guide, circle and label the following physical regions on your map:

- Amazon Basin   - Patagonia   - Mato Grosso Plateau
- Falkland Islands   - Galapagos Islands   - Cape Horn

Map 5–4: South American Bodies of Water

South America is a realm that has been influenced by its water resources, both in terms of spatial history and also economic development. On this map, you will identify the major bodies of water found inside this realm. These are divided into groups and it is recommended to use a different colored pencil or pen for each of the names per water body category. Use p. 234 as a guide.

Major Rivers:
- Amazon River   - Orinoco River   - Paraguay River
- Sao Francisco River - Madeira River   - Negro River
- Blanco River   - Magdalena River   - Tocantins River

Saltwater Bodies Adjacent to South America:
- Atlantic Ocean   - Pacific Ocean   - Strait of Magellan
- Rio de la Plata   - Caribbean Sea

Map 5–5: Major Climate Regions of South America

South America is a realm that contains a wide variety of climates. Most of these climates contain very cold temperatures throughout the year, although there are some slightly warmer climates found in the extreme southern portions of the realm. Refer to the World Climate Map (Figure G – 8) to assist with creation of this map. The major Köppen Climate types are given below for each of the climates found in this realm. Using a different colored pen/pencil for each category, outline and then color in these climate types:

Highland Climates: *Polar Climates:*
- H – Climates   - E – Climates

Tropical Climates: *Dry Climates:*
- Af   - BWh
- Aw   - BSh
- Am   - BWk
                         - BSk

Mild Mid-Latitude Climates:
- Cfa
- Cwa
- Cfb
- Csb

Map 5–6:   South American Population Densities

South America is referred to as a 'hollow continent' due to its population being primarily located along the coastlines. Using the South American Population Distribution map on Figure 5–3 (p. 239), we see that four general population density categories exist in the South American Realm. Using a different colored pen/pencil for each category, outline and color the countries/sections of countries that fit into each of these designations. Once completed, you will see a spatial pattern of population distribution emerge across the realm.

Very Dense: The highest concentration of human settlement is found in two main regions:
  • Eastern South America along the coastlines, from Recife to Buenos Aires
  • Northwestern South America, from Quito to Caracas

Dense: This population category is found just inland of the 'Very Dense' populations, extending maybe 300 miles toward the interior of the continent from the coastline areas in Eastern and Northwestern South America.

Sparse: Located in the Central and Southern Amazon Basin region.

Very Sparse: Primarily in the Andes Mountains from Peru to Cape Horn, and also the northern Amazon Basin region.

Map 5–7:   South American Indigenous and Colonial Domains

Before European settlement, South America was inhabited by a variety of indigenous groups including the Inca civilization. Once the Europeans arrived they brought both war and disease to the realm, decimating the native populations. As a result, these populations were forced to migrate inland to the mountain regions to escape. Over time, *mestizo* populations started to increase (refer to the textbook for this definition), and Africans were brought to the coastlines to work as slaves on plantations.

Using Figure 5–6 (p. 243), you will create a map of the ethnic diversity in South America today. Use a different colored pen/pencil for each ethnic group.

African: Located primarily on the Northern and Eastern coastlines of the continent

Mestizo: Found throughout the Northern regions of the realm, to Paraguay and then down the Western Coastline into Chile.

Amerindian: These people are located today in scattered clusters throughout the Amazon Basin, and also in the Andes Mountains from Colombia to Bolivia.

European: While people of European descent are located all throughout South America, their primary concentrations are found in the Southern regions including Argentina, Uruguay and Brazil.

Map 5–8: Agricultural Systems of South America

South America has a long history of agricultural systems, starting before the time of European exploration to present-day. The systems and methods used in agriculture differ quite significantly across the realm, and depend on both the physical landscapes and climate. Using Figure 5–5 (p. 242), you will create a map that shows the spatial distribution of agricultural regions found throughout the South America. Using a different colored pen/pencil for each, color in the regions where you will find each respective agricultural activity. A brief description is given for orientation, and note that some of these regions will overlap each other.

Soybeans:	Soybeans are found primarily in Central Brazil.
Plantations:	Located on the Northern and Eastern shorelines of South America
Agroforestry and 'Shifting Cultivation':	Otherwise known as 'Slash and Burn Agriculture', this activity is found throughout the Amazon Basin.
Grain Crops:	This is concentrated in Northern and Central Argentina.
Pastoralism:	Also known as grazing, this activity is mainly confined to Southern Argentina.
Dry Farming:	These are crops not requiring irrigation, and located in the southern parts of Chile. This is where many of the fruits are grown for North America during our winter season.
Nonagricultural Regions:	The primary locations where agriculture is not widely found are located in the high elevations of the Andes Mountains, from Peru to Chile.

Map 5–9: Environmental Issues in South America

The South American Realm has witnessed a substantial rise in manufacturing jobs and a move to the cities by people who formerly lived in the countryside and worked in agriculture. As a result of this movement and also heavy mining in certain locations, the realm is facing moderate to severe environmental issues in certain locations. These environmental issues are divided into categories below. Using a different colored pen/pencil for each, circle or highlight the areas and label them using their respective environmental situation.

Coastal Pollution:	With increased populations and industrial jobs locating in the major eastern cities, the eastern coastline of South America has seen a significant rise in coastal pollution problems. The Northern coastline from Colombia to Guyana also has moderate problems with coastal pollution, due to oil mining and shipping traffic.
River Pollution:	Most of South America's major rivers have some form of water quality issues, stemming from either the dumping of manufacturing waste directly into these, or from fertilizer and pesticide runoff from agriculture. These include the Amazon River and all of its tributaries, the Orinoco River, and the Paraguay River.

Air Quality Issues:	As a result of the heavy manufacturing and industry presence in the South American Realm along with very heavy urban population concentrations, poor urban air quality has become an issue in the realm. The main locations are found on the eastern shoreline, from Recife to Buenos Aires.
Deforestation:	The Amazon Basin has been a worldwide example of deforestation for decades, as migrants move to the Amazon Forest and cut down the trees to make room for farmlands and/or pastureland. The Brazilian Highlands are also heavily deforested as cities in that region expand and require more land area for their people.

Map 5–10:   Regions of the South American Realm

Using Figure 5–4 (p. 240) in the textbook, color the following regions of the South American Realm and use a different colored pen/pencil for each region and the country list below as a guide. After completing this exercise, consult the textbook to understand why each of these is considered a separate region of the realm.

The North:	Colombia, Venezuela, Guyana, Suriname, French Guiana
The West:	Ecuador, Peru, Bolivia
The South:	Chile, Argentina, Paraguay, Uruguay
Brazil:	Brazil is the only country in this region

SECTION 5.2:   Review Questions

In this section you will review the main concepts and terminology from the South American Realm by answering the questions provided below. Write in your answers in the space provided, and refer to the respective sections/pages of the chapter for the correct information.

Review Section 5.2.1   South America's Major Geographic Qualities (p. 236)

Summarize the seven Major Geographic Qualities of South America in your own words:

1.)

2.)

3.)

4.)

5.)

6.)

7.)

Review Section 5.2.2   Defining the Realm (pp. 236–237)

1.) Describe the geographic location of South America relative to the rest of the Western Hemisphere. What makes the shape of the continent unique?

2.) Give examples of why South America is a 'realm in dramatic transition':

3.) What is the role of the United States in the realm of South America today?

Review Section 5.2.3   States Ancient and Modern (pp. 237–242)

1.) Where were the Amerindians located in the realm? What are 'Altiplanos' and how did these foster development and civilization by these people?

2.) Who are the Inca people, where were they primarily located and what major contributions did they make to the realm, including those that continue in modern society today?

3.) What are the major differences between historical and present-day population patterns in South America? What factors led to this spatial change?

4.) What were the impacts of European conquering of the Amerindians, both past and present? Give at least five examples from the textbook to support your answer.

5.) Who were the two primary Iberian Invaders to this realm? When did they arrive and what spatial impact did the concept of 'land alienation' have upon these people?

6.) When did Africans come to this realm? What was the primary reason?

7.) What factors kept the South American realm isolated from the rest of the world after settlement?

Review Section 5.2.4   Cultural Fragmentation (pp. 242–243)

1.) Why is South America considered a continent of 'plural societies'? What effects have commercial and subsistence agriculture had upon the realm in this way?

2.) Describe the layering of the South American cultural landscape:

Review Section 5.2.5   Economic Integration (p. 243)

1.) What is the 'Hidrovia', and why will it be economically important to some countries in the realm? Where would it be located?

2.) Name and briefly describe the four economic trading blocks in South America, and how they differ from one another:

A.)

B.)

C.)

D.)

Review Section 5.2.6   Urbanization (pp. 243–247)

1.) What is 'urbanization'? Describe the rural to urban movement in South America during the past few decades in particular.

2.) How does a cartogram help to illustrate the population of a region or realm?

3.) What is the definition of a 'Megacity'? How many of these are found in South America? List their names and respective populations.

4.) Describe the 'model' of the Latin American city, and the different sectors and zones associated. Give examples of how these zones differ from one another, yet how they can all work together in theory to form a coherent urban unit.

5.) What are favelas, who inhabits them and how are they possibly associated with the 'Zone of Disamenity'? Where might these be located in the city?

Review Section 5.2.7   Regions of the Realm (p. 247)

1.) Name the four regions of the realm, and the respective countries that belong to each:

A.)

B.)

C.)

D.)

2.) What are the top 5 largest cities in the realm in terms of population? List these and their respective populations.

Review Section 5.2.8   The North Facing the Caribbean (pp. 248–255)

1.) Describe the current geopolitical, economic and environmental situations facing the country of Colombia today. What role has cocaine played in the development of this location?

2.) How is the physical setting of Venezuela different from that of Colombia? Use the term 'Llanos' in your description.

3.) Who is Hugo Chavez, and what role has he played in the economic development of Colombia?

4.) List and describe the 'Three Guianas' in terms of their physical setting, population density, economic development and current political issues:

A.)

B.)

C.)

Review Section 5.2.9   The West Andean South America (pp. 255–261)

1.) Summarize this region physically, culturally and economically. How is it different from The North, given these facts?

2.) What are the three sub-regions of Peru? How are they vastly different from one another? Give specific examples from the textbook to support your answer.

A.)

B.)

C.)

3.) Summarize the city of Lima (Peru) in terms of location, historical settlement, population density and economy today:

4.) What was the 'Sendero Luminoso' movement in Peru, and how did it affect the country politically and socially?

5.) Describe the differences between Ecuador and Peru in terms of population distribution, economy and cultural landscapes:

6.) What is the 'Altiplano' region of Bolivia, and why is it important to the country in several ways?

7.) Discuss the political and economic reasons that Bolivia is referred to as a 'state in trouble':

8.) Where is the 'Triple Frontier', and why is it called this? What are the reasons that this location is creating geopolitical issues for the country of Paraguay?

Review Section 5.2.10   The South Mid-Latitude South America (pp. 262–269)

1.) Why is this region called the 'Southern Cone'? What countries constitute this region?

2.) Describe the Argentine Pampas in terms of its agricultural and economic impact on the region.

3.) What are the current demographic trends in Argentina, especially in terms of urbanization? How does this compare to the rest of the South American realm?

4.) Briefly summarize how Argentina has experienced a 'boom and bust' economy in recent times, and this may have altered the spatial distribution of people and economic activity in the country.

5.) How does the country of Chile differ physically from any other in the entire realm?

6.) Building on Question 5, what are the three sub-regions Chile and what physical features characterize each?

7.) Why is Chile considered to be South America's 'greatest success story' in terms of economic development? When did this take place?

8.) How is Uruguay different from the other countries of this region physically and culturally?

Review Section 5.2.11   Brazil Giant of South America (pp. 269–280)

1.) When did Brazil become a democratic government in recent times, and what has been the economic and social impacts upon the realm?

2.) Describe the diversity of Brazil's population, giving specific examples of where these people originated and approximately how many are within each group.

3.) What has been the legacy of African heritage in Brazil? When did these people first arrive, and for what reason?

4.) Why is poverty still a prominent issue in Brazil, and what steps toward development are being taken? Give specific examples from the textbook to support your answer.

5.) Name and briefly summarize the six sub-regions of Brazil, noting the geographic differences between each (physically, culturally, economically and/or environmentally):

A.)

B.)

C.)

D.)

E.)

F.)

SECTION 5.3: Blank Maps of South America

Blank maps are provided for you to utilize with the mapping questions from Section 5.1 of the Study Guide. You are being provided with one blank map for each mapping exercise given in Section 5.1. It is a good idea to make additional copies of these blank maps to use for extra practice and review.

SECTION 5.4: Student Companion Website

Additional study tools are available on the Student Companion Website at www.wiley.com/college/deblij. Features include:

- *Flashcards* offer an excellent way to practice key concepts, ideas, and terms from the text. You can review and quiz yourself on the concepts, ideas, and terms discussed in each chapter.

- *Map Quizzes* help you to master the place names for the various regions studies. Three game-formatted map activities are provided for each chapter.

- *Chapter Review Quizzes* provide immediate feedback to true/false, multiple choice, and short answer questions.

- *Audio Pronunciation* is provided for over 2000 key words and place names from the text.

- *Annotated Web Links*

- *Area and Demographic Data*

SOUTH AMERICA

Equator

Tropic of Capricorn

SOUTH AMERICA

| 0 | 400 | 800 | 1200 | 1600 Kilometers |

| 0 | 200 | 400 | 600 | 800 | 1000 Miles |

Equator

Tropic of Capricorn

SOUTH AMERICA

| 0 | 400 | 800 | 1200 | 1600 Kilometers |
| 0 | 200 | 400 | 600 | 800 | 1000 Miles |

Equator

Tropic of Capricorn

SOUTH AMERICA

| 0 | 400 | 800 | 1200 | 1600 Kilometers |

| 0 | 200 | 400 | 600 | 800 | 1000 Miles |

Equator

Tropic of Capricorn

SOUTH AMERICA

SOUTH AMERICA

SOUTH AMERICA

Equator

Tropic of Capricorn

CHAPTER #6   SUB-SAHARAN AFRICA

SECTION 6.1:   Map Creation

In this section you will create your own detailed maps of the different major physical, cultural, geopolitical and environmental attributes of the realm. Each of these maps is critically important for understanding the Major Geographic Qualities of Sub-Saharan Africa, including their spatial distribution and functionality. By creating these on your own using the blank maps provided at the end of this chapter, you will actively learn where these different features are located and how their spatial distribution affects the ways of life in each of these respective areas.

Blank Maps for Sub-Saharan Africa are provided for you in Section 6.3. Label each of your maps with the appropriate name and title. For instance, the first map you will create will be called 'Map 6–1: Countries of Sub-Saharan Africa'.

Map 6–1:   Countries of Sub-Saharan Africa

On your first map of this realm, label the following countries of Sub-Saharan Africa. These have been divided into three regions, and you may need to use abbreviations and/or arrows if you do not have enough space to write in the full name on the map. To assist with creation of this map, refer to p. 280.

Southern Africa Countries:

- South Africa	- Lesotho	- Swaziland	- Namibia
- Botswana	- Zimbabwe	- Mozambique	- Zambia
- Madagascar	- Malawi	- The Seychelles	- Angola

Eastern Africa Countries:

- Tanzania	- Uganda	- Rwanda	- Burundi
- Somalia	- Ethiopia	- Eritrea	- Djibouti
- Sudan	- The Congo (DRC)	- Central African Republic	

Western Africa Countries:

- Congo	- Gabon	- Equatorial Guinea	- Cameroon
- Sao Tome & Principe	- Nigeria	- Chad	- Niger
- Benin	- Burkina Faso	- Togo	- Ghana
- Ivory Coast	- Liberia	- Sierra Leone	- Guinea
- Guinea-Bissau	- Gambia	- Senegal	- Cape Verde Islands

Map 6–2:   Major Cities of Sub-Saharan Africa

On this map, label the following major cities of the Sub-Saharan Africa Realm. You may need to use abbreviations and/or arrows if you do not have enough space to write in the full name on the map. To assist with creation of this map, refer to the overall map of Sub-Saharan Africa found on p. 280. Write in the city name using a different colored pencil or pen for each of the regions these belong.

Southern Africa Cities:

- Cape Town	- Port Elizabeth	- Johannesburg	- Pretoria
- Maseru	- Maputo	- Harare	- Windhoek
- Gaborone	- Lusaka	- Antananarivo	- Lilongwe
- Luanda	- Huambo	- Bulawayo	

Eastern Africa Cities:
- Dar Es Salaam   - Bujumbura   - Kigali   - Kampala
- Kinshasa   - Lubumbashi   - Nairobi   - Mogadishu
- Addis Ababa   - Djibouti   - Asmara   - Kisangani

Western Africa Cities:
- Yaoundé   - Brazzaville   - Bangui   - Libreville
- Abuja   - Lagos   - Porto-Novo   - Ouagadougou
- Abidjan   - Monrovia   - Freetown   - Conakry
- Bissau   - Dakar   - Praia

Map 6–3: *Physical Features of Sub-Saharan Africa*

Sub-Saharan Africa has a wide diversity of physical features throughout the realm that have played an important role in settlement patterns, political boundaries, distribution of mineral resources (especially diamonds), and environmental issues. For this map, you will draw in the physical features listed below. These have been divided for you into categories, and it is recommended to use a different colored pencil or pen for each of the names per category. To assist with creation of this map, refer to both the overall map of Sub-Saharan Africa found on p. 280, and also Figure 6–2, p. 284.

Mountain Ranges: Draw in the locations of these mountain ranges on your map and label them accordingly. Use a Λ symbol to denote the locations of these features.
- Highveld   - Madagascar Range   - Ethiopian Highlands
- Adamawa Highlands   - Tibesti Mountains   - Futa Jallon Highlands
- Great Karoo Cape Ranges   - Ahaggar Mountains   - Mt. Kilimanjaro

Basins:
- Kalahari Basin   - Congo Basin   - Sudan Basin
- Chad Basin   - Djouf Basin

Plateaus and Deserts:
- Bihe Plateau   - Jos Plateau   - Namib Desert

Map 6–4: *Sub-Saharan Africa Bodies of Water*

Sub-Saharan Africa is a realm that has been influenced by its water resources, both in terms of spatial history and also economic development. On this map, you will identify the major bodies of water found inside this realm. These are divided into groups and it is recommended to use a different colored pencil or pen for each of the names per water body category. Use p. 280 of the textbook as a guide.

Major Rivers:
- Congo River   - Orange River   - Niger River   - Zambezi River
- Cuanza River   - White Nile River   - Blue Nile River   - Senegal River

Inland Lakes:
- Lake Victoria   - Lake Tanganyika   - Lake Nyasa   - Lake Malawi
- Lake Chad   - Lake Kariba   - Lake Turkana   - Lake Tana

Saltwater Bodies:
- Atlantic Ocean   - Indian Ocean   - Mozambique Channel
- Gulf of Guinea   - Gulf of Adan   - Red Sea

Map 6–5:   Major Climate Regions of Sub-Saharan Africa

Sub-Saharan Africa is a realm that contains a wide variety of climates. Most of these climates contain very cold temperatures throughout the year, although there are some slightly warmer climates found in the extreme southern portions of the realm. Use the World Climates Map (Figure G – 8) of the Introductory Section as a guide.

The major Köppen Climate types are given below for each of the climates found in this realm. Using a different colored pen/pencil for each category, outline and then color in these climate types:

Tropical Climates:
- Am
- Af
- Aw

Dry Climates:
- BWh
- BSh
- BSk

Mild Mid – Latitude Climates:
- Cwa
- Cfa
- Cfb
- Csb

High Climates:
- H

Map 6–6:   Sub-Saharan Africa Population Densities

Using the Sub-Saharan Africa Population Distribution map on Figure 6–4 (p. 287), we see that four general population density categories exist in the realm. Using a different colored pen/pencil for each category, outline and color the countries/sections of countries that fit into each of these designations.

Very Dense: The highest population densities are found in two main clusters:
   • Western Africa, from Nigeria to Senegal
   • Eastern Africa, from Ethiopia to Malawi

Dense: This population category accounts for Central Africa and the eastern coastline of Africa, including Madagascar.

Sparse: Found on the western coastline from Cameroon to the Angola/Namibia border.

Very Sparse: Two main locations for the lowest population densities in the realm:
   • Southwest Africa, centered on Namibia
   • North-Central Africa, in the Sahara Desert

Map 6–7: *Spatial History: Colonial Influences in the Realm*

Using Figure 6–9 (p. 299), you will create a map that shows the colonial history and European presence found in the Sub-Saharan Africa Realm. As you can see from this map, Europe had a substantial influence in settlement of the realm from the late 1800's into the mid-1900's. Reasons for settlement included extensive forest and mineral resources, including diamonds and coal in particular.

In this map, we will examine the year 1910, in which the European presence in the realm was near its peak. Use a different colored pen or pencil and color in the regions listed below that belong with the respective country of Europe who occupied those lands during this time.

France: Most of Western Africa (note the exceptions on the map), and also Madagascar.

Britain: Three main 'clusters' of land territory (examples of countries listed below):
                   • North – Central Africa (Sudan, Uganda, Ethiopia, and part of Somalia)
                   • Western Africa (Nigeria, Ghana, Sierra Leone)
                   • Southern Africa (Zambia, Zimbabwe, Malawi, Botswana, South Africa)

Germany: Countries occupied included Cameroon, Tanzania, Benin and Namibia.

Italy: Countries occupied included Somalia and Eritrea.

Portugal: Countries occupied included Angola and Mozambique.

Belgium: Countries occupied included The Congo, Rwanda and Burundi.

Map 6–8: *Islamic Presence in Sub – Saharan Africa and The Muslim Transition Zone*

Part I: Using Figure 6–19 (p. 334), you will create a map that shows the spatial distribution of the Islamic religion in the realm, as a function of the percentage of Muslims per country. These have been divided into five categories based on percentage of people who are Muslim, and you will use a different colored pen/pencil for each category to visually determine the spatial differences in the Islamic presence.

Category 1:   Countries with Muslims as less than 5% of their population

Category 2:   Countries with Muslims as 5–10% of their population

Category 3:   Countries with Muslims as 11–35% of their population

Category 4:   Countries with Muslims as 36–70% of their population

Category 5:   Countries with Muslims as 71–100% of their population

Part II: When completed with Part I, you will refer to the map and use a black pen to superimpose the 'Muslim Transition Zone' line on the realm, showing where the boundary is approximately located between those countries with a majority of their population practicing the Islamic religion, versus those practicing other forms of religion (mainly Tribal religions and/or Christianity).

Map 6–9:   *Language Families in Sub-Saharan Africa*

Using Figure 6–10 (p. 301) as a guide, circle and label the following language families of the realm on your map. Use a different color pen/pencil (if possible) for each area, and consult the textbook for further information on the difference of these language families and their significance in the spatial history and evolution of the realm throughout time.

Niger – Congo Family: Found throughout Western, Central and Southern Africa (see map).

Nilo – Saharan Family: Located in the North – Central part of the realm.

Khoisan Family: Found in Namibia and Botswana primarily today.

Afro – Asiatic Family: Aligned with the Sahara Desert and North Africa, reaching into Ethiopia, Somalia and Eastern Tanzania.

Indo – European Family: From British legacy, this family is found mainly in South Africa.

Malay – Polynesian Family: Located in Madagascar and The Seychelles.

Map 6–10:   *Environmental Issues in Sub-Saharan Africa*

The Sub-Saharan Africa Realm presently is facing many substantial environmental problems, owing to mineral extraction (i.e. oil, diamonds), overgrazing of grasslands, and little to no environmental regulations throughout the realm. These environmental issues are divided into categories below. Using a different colored pen/pencil for each, circle or highlight the areas and label them using their respective environmental situation.

Coastal Pollution: Almost all of the Gulf of Guinea Sub-Saharan Africa's western boundary is affected by coastal pollution, sources being manufacturing waste and oil spills that are discharged here directly. The coastline of Southern Africa from Cape Town to Dar Es Salaam also has this issue, due to both polluted rivers discharging here combined with moderate to heavy shipping traffic.

River Pollution: Most of Sub-Saharan Africa's major rivers have some form of water quality issues. In particular, the Niger, Congo and Orange Rivers are the most affected by either agricultural runoff or direct dumping of pollution into these water bodies. Trace these using a blue pen/pencil and label them accordingly.

Air Quality Issues: As a result of the heavy oil refining presence in the western regions of the realm such as Nigeria, air quality issues in this region are fair to poor throughout the year. In particular, countries from Nigeria to Ghana are affected the most.

Desertification: As you progress to the Sahara Desert from Central Africa, the vegetation changes from rainforest to savannas, grassland and finally desert. From Senegal to Ethiopia, farmers have overgrazed the grasslands in a region known as the 'Sahel' and resulted in substantial conversion of these grasslands to desert as a function of the grasslands not being given time to replenish.

Soil Salinization: The overuse of soils combined with prolonged drought in some regions has led to significant soil degradation issues. These are mainly found in Central Africa from the northern part of The Congo to Nigeria and east to Ethiopia and Somalia.

Map 6–11: **Regions of the Sub-Saharan Africa Realm**

Using Figure 6–11 (p. 303) in the textbook, color the following regions of the Sub-Saharan Africa Realm and use a different colored pen/pencil for each region and the country list below as a guide. After completing this exercise, consult the textbook to understand why each of these is considered a separate region of the realm.

Southern Africa:
- South Africa   - Botswana   - Namibia   - Lesotho
- Swaziland   - Zambia   - Zimbabwe   - Malawi
- Mozambique   - Angola   - Madagascar

Eastern Africa:
- Tanzania   - Uganda   - Rwanda   - Burundi
- Kenya   - Ethiopia   - Somalia   - Djibouti
- Eritrea   - Sudan

Equatorial Africa:
- The Congo   - Congo   - Gabon   - Equatorial Guinea
- Cameroon   - C. African Republic

West Africa:
- Chad   - Niger   - Nigeria   - Benin
- Togo   - Ghana   - Ivory Coast   - Liberia
- Sierra Leone   - Guinea   - Guinea-Bissau   - Gambia
- Burkina Faso   - Mali   - Senegal

SECTION 6.2:   Review Questions

In this section you will review the main concepts and terminology from the Sub-Saharan Africa Realm by answering the questions provided below. Write in your answers in the space provided, and refer to the respective sections/pages of the chapter for the correct information.

Review Section 6.2.1   Sub-Saharan Africa's Major Geographic Qualities (p. 283)

Summarize the ten Major Geographic Qualities of Sub-Saharan Africa in your own words:

1.)

2.)

3.)

4.)

5.)

6.)

7.)

8.)

9.)

10.)

Review Section 6.2.2   Introduction to the Realm (p. 282)

1.) What effect did this realm have upon human evolution in the world? How have migrations been a crucial part of this process?

2.) What is meant by the term 'Africanization'? How has Africa been a cultural leader of the world since the first evolution of humans?

Review Section 6.2.3   Africa's Physiography (pp. 283–285)

1.) How large of a space does Africa's land area occupy?

2.) Where are the 'Bulge' and 'Horn' locations, and what countries do they occupy?

3.) How is Africa's Physiography unique from that of other continents on Earth?

4.) Where are the 'Great Lakes' of Africa? What are their names?

5.) Why could Africa be called 'The Plateau Continent'? Name some of the plateaus and their respective locations in the realm.

6.) How does the Theory of Plate Tectonics help explain Africa's physiography today?

Review Section 6.2.4   Natural Environments (pp. 286–287)

1.) Describe the symmetry of Africa's climates, and how this climate distribution accounts for the distribution of natural environments in the realm.

2.) Africa is often considered the primary location of Earth for both Savanna and Steppe vegetation. Describe each of these regions in terms of location and vegetation types.

3.) What effect did European colonization have upon the natural environments of the realm? Give at least three examples of their impact.

4.) What factors would allow Africa to have good farming potential? Why is farming difficult in the realm today, given these attributes?

Review Section 6.2.5   Environment and Health (pp. 287–289)

1.) What is the field of Medical Geography, and how does it apply to this realm?

2.) Sub – Saharan Africa is known to contain some of the highest disease rates in the world, both currently and historically. Give examples of the kinds of diseases facing this realm, and differentiate between the terms 'pandemic' and 'endemic' in your discussion.

Review Section 6.2.6   Land and Farming (pp. 289–292)

1.) What is the concept of 'land tenure', and how is it different in Africa versus European and American traditions?

2.) What effect did 'land alienation' have upon the land tenure in Africa? Give examples to support your answer.

3.) From the discussion box 'AIDS in Sub – Saharan Africa', summarize the realm's situation with this disease, in terms of spatial extent and effect upon the people in the realm. What solutions can you offer to help this problem?

4.) How has rapid population growth impacted land tenure in the realm?

5.) Describe the types of farming in the realm today, and examples of the crops grown in each. Use the terms 'intercropping', 'compound cropping', and 'diversification' in your discussion.

6.) What is the 'Green Revolution' and how could it impact Africa? Has it happened yet?

Review Section 6.2.7   Africa's Historical Geography (pp. 292–300)

1.) Why is Africa considered the 'Cradle of Mankind'? How long have humans (homo sapiens) been estimated to have existed in the realm?

2.) Why is very little known about humans and their cultures in this realm from 5,000 to about 500 years ago? Give examples that support your answer.

3.) Who were the Nok people, and what contributions did they make to African culture? When did they exist, and in what locations?

4.) How did the idea of 'regional complementarily' help develop the realm? What regions of the realm experienced development first?

5.) Why did the cultural and political focus of the realm shift over time from west to east? Which religion may have assisted in this movement?

6.) Describe the process of 'state formation' and how it helped shape the realm over time.

7.) Who were the Bantu people? Where did they live and why was their migration so geographically significant?

8.) The Slave Trade was one of the worst and darkest periods in all of recorded human history. Where were slaves taken from this realm, who was primarily responsible, and how many people were estimated to have been taken to the Americas as slaves?

9.) Why did Europeans colonize the realm in the 1700's and 1800's? Which European countries were the primary colonizers and where in the realm did they claim territory?

10.) What legacies of colonization did the Europeans leave upon the realm?

Review Section 6.2.8   Cultural Patterns (pp. 300–302)

1.) What is a 'Lingua Franca'? Why would it be especially valuable in this realm?

2.) How many languages are estimated to be spoken by people in Sub – Saharan Africa today? Name at least four of the dominant language families in the realm today:

A.)

B.)

C.)

D.)

3.) Name at least five languages that are spoken by more than 1 million people or more in Africa:

4.) How has multilingualism been a powerful force in African culture?

5.) Besides native tribal religions, what two religions are the most prominent in Africa today? Where did they spatially settle in the realm, and during what time period?

Review Section 6.2.9   The Modern Map and Traditional Society (pp. 302–304)

1.) Give at least two example of Supranationalism in the realm since the 1960's, and the names of the specific treaties or pacts that were made by member nations.

A.)

B.)

2.) How many people are estimated to live in cities in Sub – Saharan Africa? What percentage of the population does this comprise?

3.) Describe the differences between the formal and informal sectors of a city in this realm:

Review Section 6.2.10   Sub-Saharan Africa's Economic Problems (p. 304)

1.) What are the pros and cons of globalization in the realm?

2.) What is the 'crucial problem' facing Africa's economy in the decades to come? Why do you think this is such a serious threat to the development of the realm?

3.) Summarize the 'Points to Ponder' in your own words, explaining why each is a present geopolitical and/or environmental problem in the realm:

A.)

B.)

C.)

D.)

Review Section 6.2.11   Regions of the Realm (p. 305)

1.) Name the five regions of the realm, describe where they are located and identify the countries belonging to each region (for overlap, name the country in both regions):

A.)

B.)

C.)

D.)

E.)

Review Section 6.2.12   Southern Africa (pp. 305–315)

1.) Give at least three reasons why Southern Africa is considered 'Africa's richest region':

2.) What is apartheid, and what influence did it have upon the realm? Where was it primarily located?

3.) Who were the Boers, and how do they relate to a group of people called the 'Afrikaners'? What European countries do these groups represent?

4.) What effect did the 1994 election have upon the political and cultural geography of South Africa? Who was elected leader?

5.) Describe the economy of South Africa, particularly in terms of its mineral resources:

6.) Name the four countries that are considered part of the 'Middle Tier' of Southern Africa, and briefly describe each:

A.)

B.)

C.)

D.)

7.) Why is the country of Zimbabwe considered a 'tragedy' in this textbook?

8.) Name and briefly describe the five countries of the 'Northern Tier' of the region:

A.)

B.)

C.)

D.)

E.)

Review Section 6.2.13   East Africa (pp. 315–322)

1.) Describe some of the climatic differences between this region and the rest of the realm, and the effect upon the people of East Africa:

2.) What are 'rift valleys'? How have these affected the physical geography of the region?

3.) From the discussion on p. 316 ('Distinctive Madagascar'), describe how this country is very different from the rest of Africa in terms of agriculture, settlement and economy:

4.) What is the Lingua Franca of East Africa?

5.) Describe the six countries of East Africa in terms of settlement, economic geography and current geopolitical issues:

A.)

B.)

C.)

D.)

E.)

F.)

Review Section 6.2.14   Equatorial Africa (pp. 322–326)

1.) Name the eight countries that comprise this realm, and briefly describe the attributes of each:

A.)

B.)

C.)

D.)

E.)

F.)

G.)

H.)

2.) Describe the physical setting of Equatorial Africa in terms of its landscapes and climate:

3.) What political issues have faced The Congo in recent years? What effects have these had upon the people of the realm, especially in terms of migration?

Review Section 6.2.15   West Africa (pp. 326–333)

1.) Name the thirteen countries of this realm, and divide them between those with historical British versus French control:

2.) Why is this region considered to have significant 'cultural and historical momentum'? Give at least four examples from the textbook to support your answer.

   A.)

   B.)

   C.)

   D.)

3.) Describe Nigeria in terms of population, language distribution and mineral resources. How have each of these contributed to geopolitical issues inside of the country?

4.) What is the 'Middle Belt Transition Zone' of Nigeria? Where is it located?

5.) What is desertification, and how has it affected the region? Where?

Review Section 6.2.16   The African Transition Zone (pp. 333–337)

1.) Where is the African Transition Zone located? What countries does it comprise?

2.) What is the 'Islamic Front', and how is it relevant within this region?

3.) Describe the spatial history of Eritrea and Djibouti, and how they relate to Ethiopia:

4.) Why is Somalia considered a 'failed state'? How has the country been spatially divided?

SECTION 6.3: Blank Maps of Sub-Saharan Africa

Blank maps are provided for you to utilize with the mapping questions from Section 6.1 of the Study Guide. You are being provided with one blank map for each mapping exercise given in Section 6.1. It is a good idea to make additional copies of these blank maps to use for extra practice and review.

SECTION 6.4: Student Companion Website

Additional study tools are available on the Student Companion Website at www.wiley.com/college/deblij. Features include:

- *Flashcards* offer an excellent way to practice key concepts, ideas, and terms from the text. You can review and quiz yourself on the concepts, ideas, and terms discussed in each chapter.

- *Map Quizzes* help you to master the place names for the various regions studies. Three game-formatted map activities are provided for each chapter.

- *Chapter Review Quizzes* provide immediate feedback to true/false, multiple choice, and short answer questions.

- *Audio Pronunciation* is provided for over 2000 key words and place names from the text.

- *Annotated Web Links*

- *Area and Demographic Data*

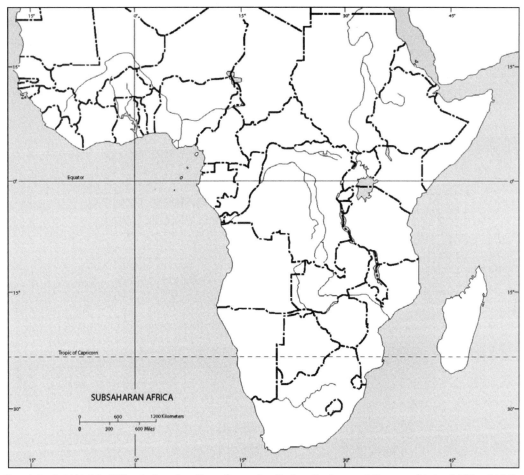

SUBSAHARAN AFRICA

Equator

Tropic of Capricorn

0   600   1200 Kilometers

0   300   600 Miles

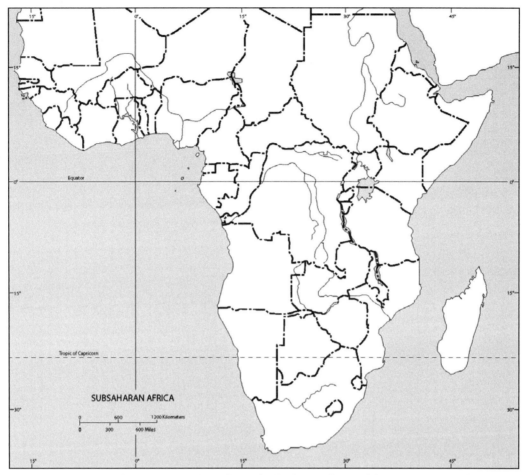

SUBSAHARAN AFRICA

0   600   1200 Kilometers
0   300   600 Miles

Equator

Tropic of Capricorn

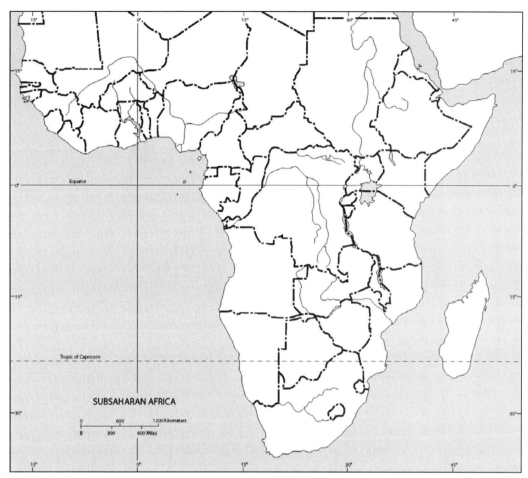

SUBSAHARAN AFRICA

Equator

Tropic of Capricorn

0   600   1200 Kilometers
0   300   600 Miles

Copyright © 2010 John Wiley & Sons, Inc. From Geography: Realms, Regions, and Concepts, 14e by deBlij and Muller

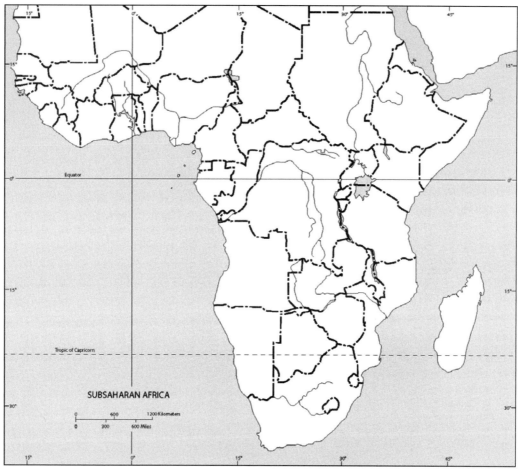

SUBSAHARAN AFRICA

Equator

Tropic of Capricorn

0   600   1200 Kilometers
0   300   600 Miles

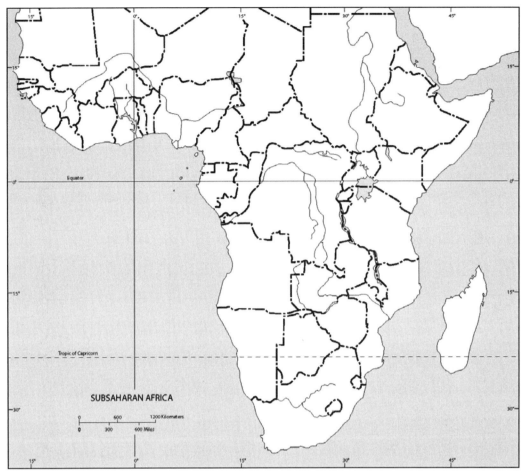

SUBSAHARAN AFRICA

Equator

Tropic of Capricorn

15° 30° 45°

0 600 1200 Kilometers
0 300 600 Miles

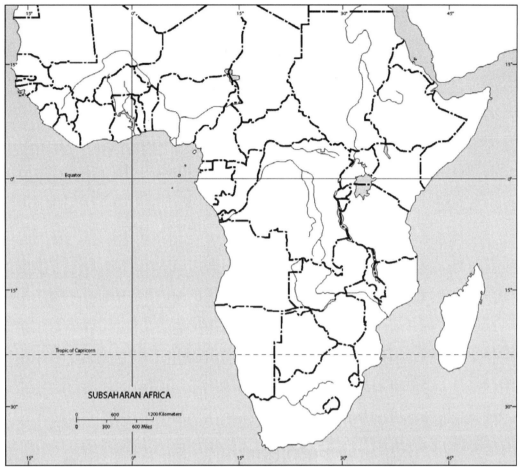

SUBSAHARAN AFRICA

Equator

Tropic of Capricorn

0   600   1200 Kilometers
0   300   600 Miles

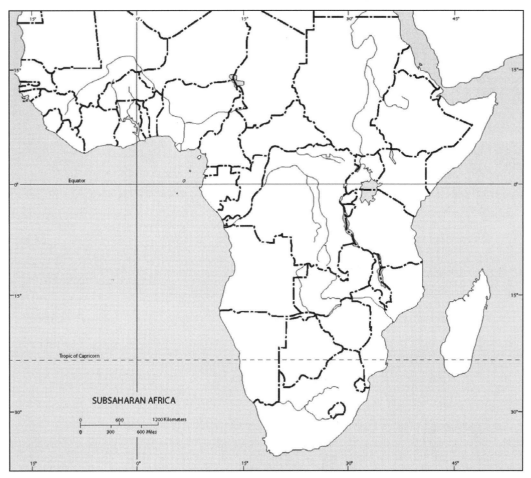

SUBSAHARAN AFRICA

Equator

Tropic of Capricorn

0   600   1200 Kilometers
0   300   600 Miles

Copyright © 2010 John Wiley & Sons, Inc. From Geography: Realms, Regions, and Concepts, 14e by deBlij and Muller

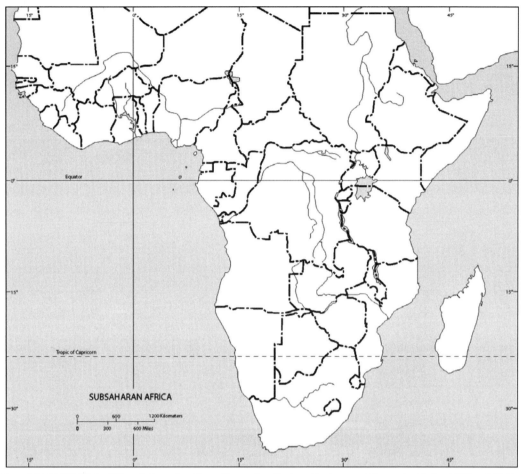

SUBSAHARAN AFRICA

Equator

Tropic of Capricorn

CHAPTER #7   NORTH AFRICA AND SOUTHWEST ASIA

SECTION 7.1:   Map Creation

In this section you will create your own detailed maps of the different major physical, cultural, geopolitical and environmental attributes of the realm. Each of these maps is critically important for understanding the Major Geographic Qualities of North Africa and Southwest Asia, including their spatial distribution and functionality. By creating these on your own using the blank maps provided at the end of this chapter, you will actively learn where these different features are located and how their spatial distribution affects the ways of life in each of these respective areas. Blank Maps for North Africa and Southwest Asia are provided for you in Section 7.3. Label each of your maps with the appropriate name and title. For instance, the first map you will create will be called 'Map 7–1: Countries of North Africa and Southwest Asia'.

Map 7–1:   Countries of North Africa and Southwest Asia

On the first map of the realm, label the following countries of North Africa and Southwest Asia. These have been divided into three regions, and you may need to use abbreviations and/or arrows if you do not have enough space to write in the full name on the map. To assist with creation of this map, refer to pp. 338–339.

Countries in North Africa:
- Morocco
- Libya
- Tunisia
- Egypt
- Sudan
- Chad
- Mauritania
- Mali
- Niger
- Western Sahara

Countries in Southwest Asia:
- Saudi Arabia
- Qatar
- U.A.E.
- Bahrain
- Oman
- Yemen
- Jordan
- Israel
- Lebanon
- Syria
- Turkey
- Iraq
- Iran
- Kuwait

Map 7–2:   Major Cities of North Africa and Southwest Asia

On this map, label the following major cities of the North Africa and Southwest Asia Realm. You may need to use abbreviations and/or arrows if you do not have enough space to write in the full name on the map. To assist with creation of this map, refer to the overall map of North Africa and Southwest Asia found on pp. 338–339.

Major Cities in North Africa:
- Cairo
- Alexandria
- Tripoli
- Khartoum
- Algiers
- Rabat
- Casablanca
- Tunis

Major Cities in Southwest Asia:
- Tel Aviv
- Jerusalem
- Amman
- Damascus
- Beirut
- Istanbul
- Ankara
- Baghdad
- Mosul
- Basra
- Kuwait City
- Riyadh
- Dubai
- Abu Dhabi
- Doha
- Mecca
- Al-Mukalla
- Muscat
- Tehran
- Shiraz

Map 7–3:   Physical Features of North Africa and Southwest Asia

North Africa and Southwest Asia are dominated by deserts, though do have some spatial diversity in terms of physical landscapes such as mountain chains and plateaus. For this map, you will draw in the physical features listed below. To assist with creation of this map, refer to both the overall map of North Africa and Southwest Asia found on pp. 338–339.

Mountain Ranges: Using a blue pen/pencil, draw in the locations of these mountain ranges on your map and label them accordingly. Use a Λ symbol to denote the locations of these features.
- Atlas Mountains   - Tibesti Mountains   - Ahaggar Mountains
- Zagros Mountains   - Elburz Mountains   - Caucasus Mountains

General Physical Features: Label these features accordingly, using a different colored pen/pencil for each type of feature listed.
- Sahara Desert   - Libyan Desert   - Run Ah – Khali (Desert)
- Western Desert   - Nubian Desert   - Syrian Desert
- Salt Desert   - Iranian Plateau   - El Djouf Basin

Map 7–4:   North Africa and Southwest Asia Bodies of Water

North Africa and Southwest Asia is a realm that is more dependent on its limited freshwater supplies than anywhere else in the world. The Nile, Tigris and Euphrates Rivers provide lifelines to the people of this realm in terms of consumption, bathing, agricultural use, and hydroelectric power. Saltwater bodies such as the Persian Gulf allow for massive quantities of petroleum to be shipped worldwide, driving the economy of the realm On this map, identify the major bodies of water found in North Africa and Southwest Asia. These are divided into groups and it is recommended to use a different colored pencil or pen for each of the names per water body category. Use pp. 338–339 as a guide.

Major Rivers:
- Nile River   - Tigris River   - Euphrates River   - Jordan River

Saltwater Bodies:
- Mediterranean Sea   - Persian Gulf   - Red Sea   - Suez Canal
- Black Sea   - Caspian Sea   - Gulf of Oman   - Arabian Sea
- Gulf of Adan   - Indian Ocean   - Atlantic Ocean   - Gulf of Sidra

Map 7–5:   Major Climate Regions of North Africa and Southwest Asia

North Africa and Southwest Asia is a realm that contains dry climates almost exclusively, being dominated by desert and grassland climates. Use the World Climate Map (Figure G – 8, Introductory Section) as a guide. The major Köppen Climate types are given below for each of the climates found in this realm. Using a different colored pen/pencil for each category, outline and then color in these climate types:

Dry Climates: *High Climates:*
- BWh   - H
- BSh
- BSk

Mild – Midlatitude Climates:
- Csa

Map 7–6: North Africa and Southwest Asia Population Densities

The human populations across the North Africa and Southwest Asia Realm are dictated by the presence or absence of water supplies. Near rivers and ports we will find very dense populations; in the deserts it will primarily be very sparse or even empty. Using the North Africa and Southwest Asia Population Distribution map on Figure 7–2 (p. 342), we see that four general population density categories exist in the North Africa and Southwest Asia Realm. Using a different colored pen/pencil for each category, outline and color the countries/sections of countries that fit into each of these designations.

Very Dense: This population category is found near the major freshwater bodies and ports. In particular, circle/color the following areas:
- The Nile River
- The Northern coastline of Africa, from Tripoli to Casablanca
- The eastern Mediterranean Sea shoreline, from Israel to Istanbul

Dense: Dense populations are found in the following locations:
- The rest of Turkey, Iraq, and Western Iran
- Western Saudi Arabia to Yemen

Sparse: These densities are located in Eastern Iran, and Southern Egypt into Sudan.

Very Sparse: All other areas in the realm, especially those in the Sahara Desert (North Africa).

Empty: The Rub Al – Khali in Saudi Arabia is entirely devoid of people

Map 7–7: Spatial History: Colonized Ottoman Provinces in the Realm

Using Figure 7–8 (p. 355), you will create a map that shows the European colonization influence in the realm, in the area referred to as the Ottoman Empire that is aligned with the Mediterranean Sea. Use a different colored pen/pencil for each country that colonized the realm, and use the descriptions below as a guide along with the textbook figure.

British: Primarily located in present-day Egypt, Sudan, Jordan, Iraq and Yemen.

France: Colonized the area in present-day Syria.

Italy: Italy settled the present-day region of north and central Libya.

Russia: Russia's influence was in extreme northern Turkey and northern Iran.

Persia: Persia was located in an area occupying present-day western Iran.

Map 7–8: The Diffusion of Islam

Using Figure 7–5 (p. 350), you will create a map that shows the diffusion of the Muslim religion from its origins in Saudi Arabia in 630 A.D. through 1600 A.D. Using the categories below and a different colored pen/pencil for each, color in the regions of Muslim expansion in the respective time frames.

Map 7–9:   Oil and Natural Gas in North Africa and Southwest Asia

The economy of North Africa and Southwest Asia is highly dependent upon oil and natural gas production, though the distribution of these mineral deposits are found in less than half of the realm's countries. Using Figure 7–9 (pp. 356–357) as a guide, circle and label the following features relating to oil and natural gas production in the realm, using a different color pen/pencil (if possible) for each item.

Areas of Oil Production: Circle the areas that are shown in the textbook as oil-producing regions, using a red pen or pencil.

Natural Gas Production: Circle the areas that are shown in the textbook as natural gas-producing regions, using a green pen or pencil.

Major Oil Pipelines: Using a black pen or pencil, draw in the major oil pipelines that are found in Libya, Algeria, Iraq, Iran, Kuwait and Egypt.

Major Oil Terminals: Using a blue pen or pencil, locate the major oil terminals of the realm by placing a dot over these locations and labeling them accordingly.

Map 7–10:   Environmental Issues in North Africa and Southwest Asia

The North Africa and Southwest Asia Realm is one of the most environmentally-degraded places in the world today. This realm is one of the most advanced realms in the world today in terms of technology and emissions standards for air and water quality. However, moderate to severe environmental issues persist in certain locations. These environmental issues are divided into categories below. Using a different colored pen/pencil for each, circle or highlight the areas and label them using their respective environmental situation.

Coastal Pollution: Almost all of the Mediterranean Sea along North Africa and Southwest Asia's southern boundary is affected by coastal pollution, due to the tremendous number of oil and gas tankers that transport petroleum to the rest of the world. Additionally, the Suez Canal, Red Sea and Persian Gulf have substantial coastal pollution issues for the same reason.

River Pollution: All of the major rivers in the realm (Nile, Jordan, Tigris and Euphrates) have moderate to severe problems with contamination, for reasons ranging from agricultural runoff to byproducts of oil and gas production being directly discharged into these water bodies.

Air Quality Issues: Poor urban air quality has been an issue in the realm in the last century as a result of the heavy oil refining presence in the North Africa and Southwest Asia Realm, along with exploding urban populations in locations such as Alexandria and Cairo. In particular, the regions from Egypt to Saudi Arabia and Iraq tend to have the worst problems with air pollution.

Soil Salinization: Regions near the major rivers such as the Nile, Tigris and Euphrates are experiencing significant issues with soil Salinization, as overusing the land for agriculture and poor irrigation techniques are creating salt buildup in the soils.

Map 7–11:   *Regions of the North Africa and Southwest Asia Realm*

Using Figure 7–10 (pp. 362–363) in the textbook, color the following regions of the North Africa and Southwest Asia Realm and use a different colored pen/pencil for each region and the country list below as a guide. After completing this exercise, consult the textbook to understand why each of these is considered a separate region of the realm.

- The Maghreb and its Neighbors   - The Middle East
- Egypt and the Lower Nile Basin   - The Empire States
- The Arabian Peninsula   - Turkestan

SECTION 7.2:   Review Questions

In this section you will review the main concepts and terminology from the North Africa and Southwest Asia Realm by answering the questions provided below. Write in your answers in the space provided, and refer to the respective sections/pages of the chapter for the correct information.

Review Section 7.2.1   Major Geographic Qualities (p. 341)

Summarize the ten Major Geographic Qualities of North Africa and Southwest Asia in your own words:

1.)

2.)

3.)

4.)

5.)

6.)

7.)

8.)

9.)

10.)

11.)

Review Section 7.2.2   Defining the Realm (pp. 341–344)

1.) Explain how the 'centrality' of the realm has affected settlement and cultural patterns over time.

2.) How many people are estimated to live here? Where are most of the populations clustered? Be specific and name exact locations.

3.) There are three monikers for this realm, each of which is geographically-confusing but can be considered relevant. These are listed below; describe why each term could be appropriate for this location:

A.) The 'Middle East':

B.) An 'Arab World':

C.) An 'Islamic World':

4.) How many states and territories are located here? Which seem to be the most important, and why?

Review Section 7.2.3   Hearths of Culture (pp. 344–347)

1.) This realm has seen four major historical events throughout the course of human history. Describe the events that occurred in the following time periods, and why they are relevant to both the realm and to human history:

A.) 2 Million Years Ago:

B.) 100,000 Years Ago:

C.) 10,000 Years Ago

D.) 1,000 Years Ago

2.) What do the terms 'Cultural Environment' and 'Cultural Ecology' mean, and why are they relevant in this realm of the world?

3.) Where is 'Mesopotamia' located, what does the term mean and why is this region so important in terms of human history? Give specific examples from the textbook to support your answer.

4.) Where is the 'Fertile Crescent' located? Name specific countries found within this region, and describe why this area was historically significant for development in this realm.

5.) How could climate change have substantially altered the human landscapes in this realm in recent times? Are there continuing examples of this problem today found here?

Review Section 7.2.4   Stage for Islam (pp. 347–350)

1.) Describe the beginnings of Islam, starting in the year 611 A.D. Why is Mecca considered so important in this religion?

2.) How did Islam help to bind the realm together? Give at least four examples from the textbook.

3.) What are the Five Pillars of Islam?

    A.)

    B.)

    C.)

    D.)

    E.)

4.) Define the five types of diffusion below, and briefly describe how each promoted the spread of Islam:

    A.)

    B.)

    C.)

    D.)

    E.)

5.) Describe the spread of Islam in terms of where it is primarily located in the world today. How does the term 'Islamization' help to describe this process?

Review Section 7.2.5   Islam Divided (pp. 350–355)

1.) What are the two main branches of Islam, and how are they different? Which contains the highest percentage of Muslims today?

2.) What effect did 'religious revivalism' have upon further division of these two branches of Islam? What countries were primarily involved?

3.) Who is the 'Ayatollah', and how did he further promote Islam into Iran's government policies?

4.) Why did Iran and Iraq have a war, when did it occur and what were the reasons for this situation?

5.) Islamic revivalist fundamentalism became an important issue among the people of this realm for what reasons in the 1990's? Give specific examples to support your answer.

6.) Militant Islamic carried out attacks against the 'Oil Guzzling Western Countries' because they found these countries guilty of what four issues in particular?

A.)

B.)

C.)

D.)

7.) Building on Question #6, name some of the attacks and in which countries they have been located, from the 1990's to present-day.

8.) According to the textbook, what were the ultimate goals of Osama (Usama) Bin Laden in forming the terrorist group Al-Qaeda? What is 'Wahhabism' and how does this term relate to this situation?

9.) From the discussion box on p. 353, after reading both sides of the argument what is your opinion regarding the status of Islam- should it be revived or reformed? Give details to support your answer.

10.) What United Nations-sponsored event in 1948 created an entirely new set of religious and political boundary issues within the realm? Where is this located, and what city lies in the middle of this battleground?

11.) When was the Ottoman Empire and where was it centered? What countries were created in this realm today from the remnants of this empire?

12.) How did European powers divide the realm in terms of political boundaries after the Ottoman Empire? Which European countries had the most influence in this realm?

Review Section 7.2.6   The Power and Peril of Oil (pp. 355–360)

1.) The developed world (i.e Europe and the USA in particular) might have left this realm alone; however, two recent events caused intervention and interest here. Name these two events and where they are primarily located:

A.)

B.)

2.) Which five countries control 77% of the world's oil supply? Name these and the amount of oil each produces per year, in billions of barrels. How does this compare to the USA production?

A.)

B.)

C.)

D.)

E.)

3.) How has the uneven distribution of oil wealth affected the political and social aspects of countries in the realm? Give examples to support your answer.

4.) Why did the oil-rich countries originally have to depend on the Western World for assistance in mining this resource? What was the fear by the local governments of this situation?

5.) What role did the USA have in Iran's political affairs and negotiations with the British from the 1950's until 1979? Why was this important in terms of oil production here?

6.) Oil and gas production in the realm have resulted in ten major geographic impacts. List and briefly describe each of these in the space provided below:

A.)

B.)

C.)

D.)

E.)

F.)

G.)

H.)

I.)

J.)

7.) Summarize the four 'points to ponder' (p. 360) in your own words, and why you think these are geographically-significant in the coming century.

Review Section 7.2.7 Regions of the Realm (pp. 360–361)

1.) Why is it considered a challenge to define regions within this realm today?

2.) Name the six regions of the realm, and list the countries that are found within each:

   A.)

   B.)

   C.)

   D.)

   E.)

   F.)

3.) What are the four largest cities in the realm today, in terms of population? List these and their respective population levels.

Review Section 7.2.8 Egypt and the Lower Nile Basin (pp. 361–368)

1.) Why is Egypt considered in a pivotal position geographically? Give examples, both physical and cultural, to support your answer.

2.) Why is the Nile River considered vital for this region? What percentage of the population lives within 20 km (12 miles) of the river?

3.) What impact did the completion of the Aswan High Dam (Egypt) have upon the region, both physically and culturally?

4.) Who are the 'Felleheen' people of Egypt, and what environmental problems are threatening their way of life today?

5.) Summarize Cairo's geographic significance physically, historically, and culturally today.

6.) Name the six sub-regions of Egypt, and the reason that these are separate from one another.

   A.)

   B.)

C.)

D.)

E.)

F.)

7.) Summarize Sudan physically and culturally, especially compared to Egypt. How is this country different in the North versus the South?

8.) Compare Sudan economically from its independence in 1956 to present-day. What allowed the transformation of its economy, and when?

9.) What happened in Darfur that became such an international issue? What events led to this problem, and how did cultural differences impact and accelerate this situation?

Review Section 7.2.9   The Maghreb and its Neighbors (pp. 368–370)

1.) How is this region physically different from the rest of the realm? Give examples.

2.) Name the five countries of this region, and summarize their main geographic qualities that make these very different from one another.

A.)

B.)

C.)

D.)

E.)

3.) What five countries are located between the Maghreb and the Sahel? Why are these collectively considered different than the 'main' five countries of this region?

Review Section 7.2.10   The Middle East: Crucible of Conflict (pp. 370–379)

1.) Define the term 'Middle East' today in terms of its geographic location and significance.

2.) Describe Iraq in terms of physical size and population today. Why did the USA invade and overthrow Saddam Hussein in 2003, and what problems have come as a result of this intervention?

3.) What have been some of the wider impacts of the War in Iraq? Who are the Kurds and how have they been affected by this intervention?

4.) Summarize Syria, Jordan and Lebanon in terms of physical, cultural and economic traits. How do these differ from one another, yet how are they alike?

5.) When was the country of Israel created, and by who? Why did this create an immediate geopolitical situation in this region?

6.) What geographic impact did the week-long 1967 War have upon Israel and its neighboring countries? Where did the Palestinians migrate, and what are their respective populations today in these locations?

7.) Describe Israel today in terms of physical size and population. Given this, why does this country maintain such importance in the region today?

8.) What are the five primary obstacles to a proposed 'two-state' division of Israel and Palestine today? Briefly describe each and why these are potential issues.

A.)

B.)

C.)

D.)

E.)

Review Section 7.2.11 *The Arabian Peninsula (pp. 379–384)*

1.) Why is this area considered its own region? Give physical and cultural examples to support your answer.

2.) Summarize Saudi Arabia in terms of cultural importance and economic significance. How does this country compare to the others in this region?

3.) Kuwait is considered strategic for both economic and physical regions, hence the reason Iraq invaded this tiny country in 1990. Give reasons why it would be considered important in these ways.

4.) List and briefly describe the remaining six countries in this region, comparing them physically, culturally and especially economically against one another.

A.)

B.)

C.)

D.)

E.)

F.)

Review Section 7.2.12   The Empire States (pp. 384–391)

1.) What are the 'Empire States', and why are they considered a separate region in this realm?

2.) What religious minorities are found in this region, and where are they located? What issues have these people created both historically and today in geopolitical issues?

3.) Describe the country of Iran in terms of its physical features and agricultural opportunities. How is this country different from others found in this realm?

4.) What are 'Qanats', and how have these helped Iran develop over time?

5.) Summarize Iran today economically and politically. What challenges face this country in the future?

Review Section 7.2.13   Turkestan (pp. 391–398)

1.) Why is this region called 'Turkestan'? What cultural differences exist here, compared to the rest of the realm?

2.) List and briefly describe the five former USSR Republics found in this region:

A.)

B.)

C.)

D.)

E.)

3.) Why has Afghanistan been a center of geopolitical issues from the late 1970's through present-day? Give specific examples from the textbook to support your answer.

SECTION 7.3:   Blank Maps of North Africa and Southwest Asia

Blank maps are provided for you to utilize with the mapping questions from Section 7.1 of the Study Guide. You are being provided with one blank map for each mapping exercise given in Section 7.1. It is a good idea to make additional copies of these blank maps to use for extra practice and review.

SECTION 7.4:   Student Companion Website

Additional study tools are available on the Student Companion Website at www.wiley.com/college/ deblij. Features include:

- *Flashcards* offer an excellent way to practice key concepts, ideas, and terms from the text. You can review and quiz yourself on the concepts, ideas, and terms discussed in each chapter.

- *Map Quizzes* help you to master the place names for the various regions studies. Three game-formatted map activities are provided for each chapter.

- *Chapter Review Quizzes* provide immediate feedback to true/false, multiple choice, and short answer questions.

- *Audio Pronunciation* is provided for over 2000 key words and place names from the text.

- *Annotated Web Links*

- *Area and Demographic Data*

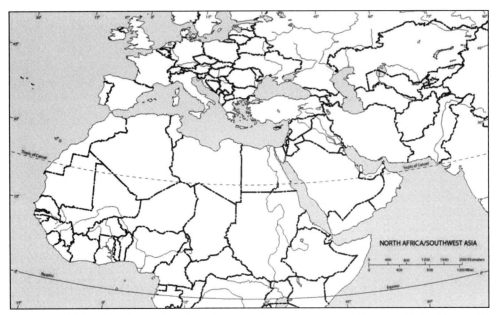

NORTH AFRICA/SOUTHWEST ASIA

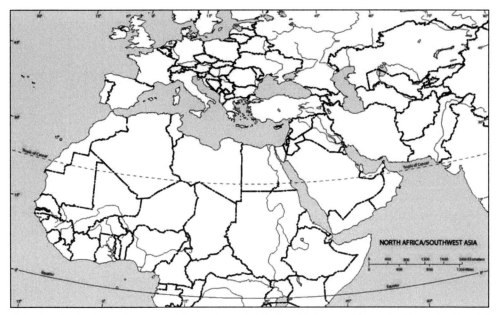

NORTH AFRICA/SOUTHWEST ASIA

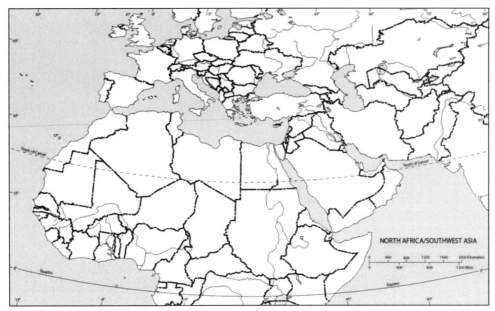

NORTH AFRICA/SOUTHWEST ASIA

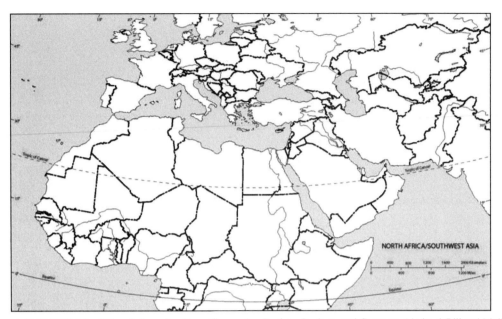

NORTH AFRICA/SOUTHWEST ASIA

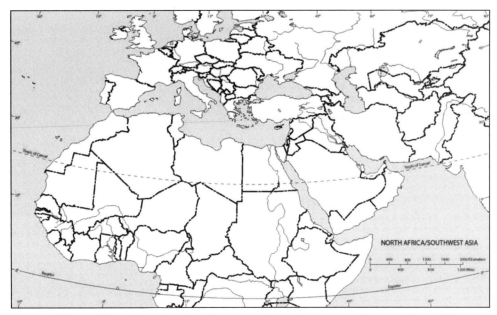

NORTH AFRICA/SOUTHWEST ASIA

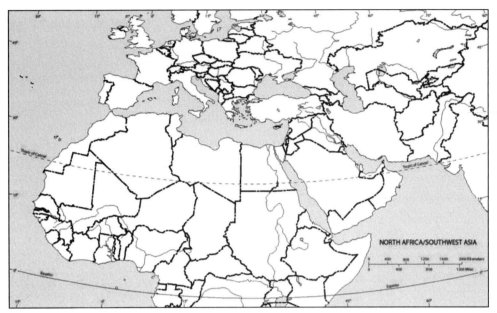

NORTH AFRICA/SOUTHWEST ASIA

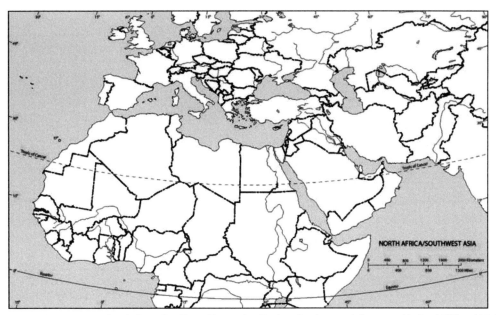

NORTH AFRICA/SOUTHWEST ASIA

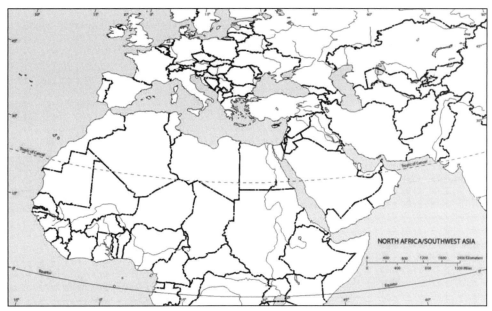

NORTH AFRICA/SOUTHWEST ASIA

CHAPTER #8   SOUTH ASIA

SECTION 8.1:   Map Creation

In this section you will create your own detailed maps of the different major physical, cultural, geopolitical and environmental attributes of the realm. Each of these maps is critically important for understanding the Major Geographic Qualities of South Asia, including their spatial distribution and functionality. By creating these on your own using the blank maps provided at the end of this chapter, you will actively learn where these different features are located and how their spatial distribution affects the ways of life in each of these respective areas. Blank Maps for South Asia are provided for you in Section 8.3. Label each of your maps with the appropriate name and title. For instance, the first map you will create will be called 'Map 8–1: Countries of South Asia'.

Map 8–1:   Countries of South Asia

On the map, label the following countries of the South Asian Realm. These have been divided into three regions, and you may need to use abbreviations and/or arrows if you do not have enough space to write in the full name on the map. To assist with creation of this map, refer to p. 400.

- India   - Pakistan   - Afghanistan   - Sri Lanka
- Nepal   - Bhutan   - Bangladesh   - The Maldives

Map 8–2:   Major Cities of South Asia

On this map, label the following major cities of the South Asian Realm. Although this is the most dense populated realm in the world, over 75% of its people live in the countryside and so there is a noticeable lack of major cities throughout the realm. To assist with creation of this map, refer to the overall map of South Asia found on p. 400. Write in the city name using a different colored pencil or pen for each of the regions these belong.

- Delhi   - Mumbai   - Bangalore   - Kolkata
- Islamabad   - Karachi   - Kabul   - Kathmandu
- Dhaka   - Colombo   - Male (Maldives) - Thimphu

Map 8–3:   Physical Features of South Asia

South Asia has a diversity of physical features throughout the realm that have played an important role in settlement patterns, political boundaries and environmental issues. For this map, you will draw in the physical features listed below. These have been divided for you into categories, and it is recommended to use a different colored pencil or pen for each of the names per category. To assist with creation of this map, refer to both the overall map of South Asia found on p. 400, and also Figure 8–4,

Mountain Ranges: Using a blue pen/pencil, draw in the locations of these mountain ranges on your map and label them accordingly. Use a Λ symbol to denote the locations of these (i.e. draw a line of upside-down triangles to show where the mountain range belongs).
- Himalayas Mountains   - Hindu Kush Mountains   - Sulaiman Range
- Vindhya Range   - Eastern Ghats   - Western Ghats

General Physical Regions: These regions all have a physical characteristic in common, whether being plateaus, coastlines or deserts. Using Figure 8-4 (p. 52) as guide, circle and label the following physical regions on your map using a green pen/pencil:

- Ganges Plain
- Chota Nagpur Plateau
- Coromandel Coast

- Thar Desert
- Konkan Coast
- Golconda Coast

- Deccan Plateau
- Malabar Coast

Map 8–4: South Asian Bodies of Water

On this map, you will identify the major bodies of water found inside and around this realm. These are divided into groups and it is recommended to use a different colored pencil or pen for each of the names per water body category. Use p. 400 as a guide.

Major Rivers:
- Ganges River
- Godavari River

- Brahmaputra River
- Krishna River

- Narmada Rive
- Indus River

Major Saltwater Bodies:
- Indian Ocean
- Gulf of Khambhar

- Bay of Bengal
- Gulf of Kutch

- Arabian Sea
- Gulf of Mannar

- Laccadive Sea
- Mouths of the Ganges

Map 8–5: Major Climate Regions of South Asia

South Asia is a realm defined by both its tropical monsoon and high climates. Deserts exist in the west, and the climates transition from tropical into mild mid-latitude as you go north, before reaching the Himalayas. Use the World Climates Map (Figure G – 8, Introductory Chapter) as a guide.

The major Köppen Climate types are given below for each of the climates found in this realm. Using a different colored pen/pencil for each category, outline and then color in these climate types:

Tropical Climates:
- Af
- Am
- Aw

Dry Climates:
- BWh
- BSh
- BSk

Mild Mid-Latitude Climates:
- Cwa

High Climates:
- H

Map 8–6: South Asian Population Densities

The human populations across the South Asian Realm are the most dense in the world, especially when you consider that India alone contains over 1.2 billion people living in in area 35% the size of the continental USA! Using the South Asian Population Distribution map on Figure 8–3 (p. 404), we see that though extremely dense populations exist here, four additional population density categories exist in the South Asian Realm. Using a different colored pen/pencil for each category, outline and color the countries/sections of countries that fit into each of these designations. Once completed, you will see a spatial pattern of population distribution emerge across the realm.

Extremely Dense:	This category of population density is unique to the South Asian Realm, and found in two locations: • In NE India and in Bangladesh, along the Ganges River • In SW India, from Mumbai to the southern tip of India
Very Dense:	These populations are found throughout the rest of India except for the Northern areas; also found in Pakistan centered on the Indus River, and in Sri Lanka.
Dense:	This population category accounts for the rest of Eastern Pakistan, as well as southern Nepal and Bhutan. Also found in The Maldives.
Sparse:	This category is a thin sliver existing between the dense populations of Nepal and Bhutan, to the northern parts of these countries where the Himalayas dominate. Sparse populations additionally are found in Afghanistan and western Pakistan.
Very Sparse:	Primarily in the Himalayas Mountains in Northern Nepal/Bhutan and Northern India.

Map 8–7:   *The Religions of South Asia*

In this map, you will show the spatial distribution of religions throughout the realm. Thought of primarily as a Hindu realm, the religious diversity in reality throughout South Asia is substantial. Use the descriptions below to guide you in creation of this map, and use different colored pens/pencils for each religion in the realm.

Buddhists:	Primarily located in Bhutan and Sri Lanka, along with far eastern India.
Muslims:	Located in Afghanistan, Pakistan, and Bangladesh.
Sikhs:	A minority religion that branched away from Hinduism and a group fiercely loyal to the British, they are identified by the turbans the males wear (men are not allowed to cut their hair). These people are found in Northwest India primarily.
Jains:	Another religion that had its roots in Hinduism, these are a merchant class that gained tremendous wealth and are responsible for some of the most massive and famous structures in India. They are located primarily in Western India, centered on Mumbai.
Parsis:	A religion composed of both Islam and Hinduism fundamentals, these people are found along the southwestern coastline of India, south of Mumbai to near Bangalore.
Christians:	Though a minority presence, Christians can be found in dense pockets in Southern India, as well as the Maldives.
Hindus:	Found throughout all of India **except** in the locations listed above. This is also the majority religion of Nepal.

Map 8–8:   The Muslim Presence in India

Using Figure 8–14 (p. 432), you will create a map that shows the spatial distribution of Muslims in an otherwise Hindu India, as of the year 2001. Religious conflict has led to significant geopolitical tension in the country and the realm as a whole, and therefore an understanding of the primary locations of religious 'mixing' between Hindus and Muslims is essential. These are divided into groups by percentage, and it is recommended to use a different colored pen/pencil for each percentage group.

India Regions with 0–4% Muslim:	Mainly in Eastern India, just below Kolkata. Also found in extreme Eastern India near China.
India Regions 5–9% Muslim:	Found throughout southeast and western India, away from the large cities of the realm.
India Regions 10–24% Muslim:	These concentrations are found in the more urban regions of southwest and northeast India.
India Regions 25–49% Muslim:	Areas with $1/4$ to $1/2$ of the population being Muslim are found along the border with Bangladesh in the east.
India Regions over 50% Muslim:	The highest population density of Muslims are found in extreme northern India, in a region known as Kashmir.

Map 8–9:   Agricultural Activity in South Asia

There are many Geopolitical Issues facing the South Asian Realm today. Many of these have a spatial component, as certain groups of minorities want independence from the respective country in which they reside. Other issues stem from land rights to the continued growth and power of the South Asian Union.

Using Figure 8–15 (p. 442) as a guide, circle and label the following areas on your map, displaying where these major Geopolitical Issues are found today. Use a different color pen/pencil (if possible) for each area, and consult the textbook for further information on the situation in these locations.

- Rice	- Wheat	- Cotton	- Millet
- Groundnut	- Shifting Cultivation	- Coconut	- Chickpea
- Corn	- Plantations		

Map 8–10:   Languages in South Asia

Using Figure 8–5 (p. 409) as a guide, circle and label the following language families of South Asia on your map. Use a different color pen/pencil (if possible) for each area, and consider the influence of the British in particular when creating this map and noticing the spatial variation of these languages.

Dravidian Language Family:	The Dravidian language family is located primarily in Southeast India, extending inland to Bangalore.
Sino – Tibetan Language Family:	This family is found in the extreme northern and far eastern regions of India, along the border of China.

Austro – Asiatic Language Family:	This language family is spoken in three small clusters centered near the eastern Ganges Plain and along the border of Bangladesh.
Malay – Polynesian Language Family:	The Malay – Polynesian family is the most sparsely spoken in the realm, and located primarily in the Maldives.
Indo – European Language Family:	This language family was brought to the realm primarily by the British and French during the initial period of world exploration in the 1500's and 1600's, and is located throughout the rest of India, both dominant in the central and northern regions.

Map 8–11: *Environmental Issues in South Asia*

The South Asian Realm is one of the worst places in the world today for environmental degradation, accelerated by the rise of global outsourcing into manufacturing jobs in the realm and a lack of environmental standards. These issues are divided into categories below. Using a different colored pen/pencil for each, circle or highlight the areas and label them using their respective environmental situation.

Sealevel Rise:	Bangladesh is highly susceptible to sealevel rise, especially as the entire country and surrounding regions in Eastern India are only 2–5 feet above sea level on average.
Coastal Pollution:	The lack of environmental regulations has resulted with industries dumping toxins directly into the open waters offshore of India. Coastal pollution applies to the shoreline of the entire country of India, along with the shoreline of Pakistan and Bangladesh.
River Pollution:	For the same reasons given above, river pollution is a big problem in this realm. Every major river has a moderate to high degree of contamination, including the Ganges, Brahmaputra, Narmada and Indus Rivers. Agricultural runoff also contributes to this situation in the realm.
Air Pollution:	With major cities averaging 10–21 million people in their metropolitan areas and extremely or very dense populations throughout the country, India is a substantial source of air pollution. In particular, circle the country but exclude the northern regions near Pakistan. Also include the country of Bangladesh as well.

Map 8–12: *Regions of the South Asian Realm*

Using Figure 8–2 (p. 403) in the textbook, color the following regions of the South Asian Realm and use a different colored pen/pencil for each region and the country list below as a guide. After completing this exercise, consult the textbook to understand why each of these is considered a separate region of the realm.

- Mainland India   - Peninsular South   - Mountainous North
- Island South   - West Region   - East Region

SECTION 8.2:   Review Questions

In this section you will review the main concepts and terminology from the South Asian Realm by answering the questions provided below. Write in your answers in the space provided, and refer to the respective sections/pages of the chapter for the correct information.

Review Section 8.2.1   South Asia's Major Geographic Qualities (p. 402)

Summarize the eleven Major Geographic Qualities of South Asia in your own words:
  1.)
  2.)
  3.)
  4.)
  5.)
  6.)
  7.)
  8.)
  9.)
 10.)
 11.)

Review Section 8.2.2   Defining the Realm (pp. 402–405)

1.) What is South Asia's major unifying force? Why this has had an impact upon the realm?

2.) Describe the size of South Asia physically, compared to the amount of population it contains.

3.) Name the seven countries of this realm.

4.) What is Kashmir, where is it located and why is it a continuing issue in the drawing of political lines between India and Pakistan?

5.) Why is Pakistan considered part of this realm instead of the North Africa and Southwest Asia realm?

Review Section 8.2.3   South Asia's Physiography (pp. 405–407)

1.) Describe the role of the mountains in this realm giving both 'life and death' to people here.

2.) What political problems are contained within these mountains, and what groups are responsible for these issues?

3.) What is the 'monsoon'? Where does it impact South Asia, and during what time of year? Using the box on p.407, briefly describe the four steps in this weather process.

4.) Name and describe the three physiographic regions of South Asia. Where is the Deccan Plateau, and why is it geographically significant?

A.)

B.)

C.)

Review Section 8.2.4   Locals and Invaders (pp. 408–412)

1.) The Ganges River basin has had what historical significance upon the realm? When did people first arrive to this area?

2.) Where is the Indus Valley region, and when did people settle these lands? What were the names of some of the groups of these people?

3.) Who were the Aryans, when did they arrive and what impact did they have upon the Indus Valley region?

4.) The Aryans brought a form of religion with them to the Indus Valley region. What was this religion, and what present-day religion did this evolve into?

5.) What is the 'Caste System' of Hinduism? How did this lead to the formation of Buddhism in the realm, and when did this occur?

6.) Using Figure 8–6 as your guide, name the top three languages spoken in the realm today and their respective populations. How does this compare to the overall estimated population of the realm (1.5 billion people)?

7.) Who were the next group of invaders to the realm after the Aryans? What was the name of their empire, and where did they primarily settle?

8.) What cultural impacts did the short-lived Gupta empire have upon South Asia?

9.) When did Muslims arrive to the realm and where did they settle? Why did millions of Hindus convert to Islam?

10.) When did the Europeans arrive to the realm? Who were the major countries involved in this expedition to South Asia?

11.) What impact did the British East India Company have upon the economy and political geography of the realm? When did this take place?

12.) Why was unifying the realm difficult for the British? What major cities in India were developed during this time of British rule?

13.) The partition of British India in 1947 resulted in what cultural and geopolitical issues? Who were the groups most active in promoting independence in the realm? Why did this cause a massive refugee movement in a short period of time?

Review Section 8.2.5   South Asia's Population Dilemma (pp. 413–417)

1.) Define the field of 'Population Geography', using the terms demography, population distribution and population density in your explanation.

2.) Briefly describe how the Demographic Transition Model shows a population explosion in the 1900's, and the reasons for this pattern.

3.) Why did population rates increased dramatically in this realm in the last 100 years? Are these slowing down, leveling off, or still rapidly increasing?

4.) Where were the highest population growth rates from 1981–2001 in India, and what factors led to these geographic patterns we see today?

5.) Describe the growth rates of Pakistan and Bangladesh in comparison to India. What are their respective populations today?

Review Section 8.2.6   South Asia's Burden of Poverty (p. 417)

1.) Summarize the problem of poverty in this realm by giving at least five facts relating to the situation:

A.)

B.)

C.)

D.)

E.)

2.) Why is poverty so severe in this realm, as seen in the textbook? Offer four reasons for this problem:

A.)

B.)

C.)

D.)

Review Section 8.2.7   The Latest Invasions (pp. 417–418)

1.) What geopolitical events in this decade have dramatically altered the political and economic status of South Asia? Give specific examples from the textbook to support your answer.

2.) Summarize the four geopolitical 'points to ponder' in your own words, and briefly state why you think each is important in the future events and development of the South Asian realm.

A.)

B.)

C.)

D.)

Review Section 8.2.8   Regions of the Realm (pp. 126–128)

1.) List the top four largest cities in this realm and their respective populations:

A.)

B.)

C.)

D.)

Review Section 8.2.9   Pakistan: On South Asia's Western Flank (pp. 419–426)

1.) Name five reasons why Pakistan is considered a region of South Asia, using examples from p. 419 of the textbook.

   A.)

   B.)

   C.)

   D.)

   E.)

2.) Why is Pakistan referred to as the 'Gift of the Indus'?

3.) After independence, why was the capital of Pakistan moved from Karachi to Islamabad? How does this location exemplify the principle of a 'forward capital'?

4.) Why does Pakistan have a long way to go in terms of democracy and geopolitical issues?

5.) Name and briefly describe the four sub-regions of Pakistan, in terms of where they are located, physical and cultural features, and current issues today.

   A.)

   B.)

   C.)

   D.)

6.) Where was 'East Pakistan' located? When did it gain its independence and for what reasons? By what name is this country known today?

Review Section 8.2.10   The Problem of Kashmir (pp. 422–423)

1.) Summarize the situation in Kashmir by listing at least five facts relating to the current political and cultural troubles within this area:

   A.)

   B.)

   C.)

   D.)

   E.)

2.) What events propelled Kashmir from a site of geopolitical trouble to one of a potential nuclear confrontation between India and Pakistan?

Review Section 8.2.11   India Astir (pp. 426–445)

1.) Give at least four examples of India's political and economic success in recent years:

A.)

B.)

C.)

D.)

2.) What impact did British colonialism have upon India? What modifications did an independent India make to their government and lifestyle after the end of British rule?

3.) Given this success, why has India remained a very poor country since its independence? Offer at least three reasons for this situation.

A.)

B.)

C.)

4.) How many states and territories comprise India? What is the country's estimated overall population?

5.) What environmental issues are forefront concerning the Ganges River? Why is this river important in terms of both agriculture and religion in India?

6.) The 'McMahon Line' and the 'Raj' have both been very politically influential in the region. Briefly describe each and its respective impact.

7.) Who are the Sikhs, and how are they different from Hindus? What is the name of the land they wanted to claim for independence from India? What provisions did India make for these people?

8.) Describe the Muslim presence in India today in terms of population, clusters of settlement, and current political issues.

9.) Name and describe the Castes in Hinduism. What percentage of the population does each constitute, and what kinds of jobs will each usually possess?

10.) What is 'Hindutva', and what political and social impact is it having upon India?

11.) What was the 'Naxalite Movement', when was it prominent in India and how is it affecting the region today?

12.) Name and briefly describe the five centripetal forces of India:

A.)

B.)

C.)

D.)

E.)

13.) How has urbanization impacted the region, and when did this movement begin? Which cities have seen the highest population gains as a result of urbanization?

14.) Describe how globalization has had a tremendous impact upon India today, giving specific examples from the textbook to support your answer.

15.) What role does farming have in the region today?

16.) How has the growth of India as a manufacturing and service center created problems with energy supplies? Give examples and offer solutions to this problem.

17.) Compare and contrast eastern India with western India, and why these areas are so different geographically in terms of populations, culture and economy.

Review Section 8.2.12   Bangladesh: Challenges Old and New (pp. 445–447)

1.) Describe Bangladesh in terms of population and economic opportunities today. Why do you think it is in this situation?

2.) What natural hazards plague the country today? When do these occur?

3.) What is the primary religion of Bangladesh? What geopolitical conflicts are currently found in the region, and do these relate to religious differences in the realm? If so, why?

Review Section 8.2.13   The Mountainous North (pp. 447–448)

1.) Compare and contrast the countries of Nepal and Bhutan in terms of physical features, cultures and religion, and economic opportunity. What similarities and differences do you notice?

2.) Why is the country of Nepal considered a 'failed state'? What factors led to this situation?

3.) Bhutan is referred to as a 'buffer state' for what primary geographic reasons?

Review Section 8.2.14   The Southern Islands (pp. 448–452)

1.) Name and briefly describe the two countries that constitute this region.

   A.)

   B.)

2.) Why are these countries considered very similar yet vastly different? Use political, economic, religious and physical examples to support your answer.

SECTION 8.3:   Blank Maps of South Asia

Blank maps are provided for you to utilize with the mapping questions from Section 8.1 of the Study Guide. You are being provided with one blank map for each mapping exercise given in Section 8.1. It is a good idea to make additional copies of these blank maps to use for extra practice and review.

SECTION 8.4:   Student Companion Website

Additional study tools are available on the Student Companion Website at www.wiley.com/college/deblij. Features include:

• *Flashcards* offer an excellent way to practice key concepts, ideas, and terms from the text. You can review and quiz yourself on the concepts, ideas, and terms discussed in each chapter.

• *Map Quizzes* help you to master the place names for the various regions studies. Three game-formatted map activities are provided for each chapter.

• *Chapter Review Quizzes* provide immediate feedback to true/false, multiple choice, and short answer questions.

• *Audio Pronunciation* is provided for over 2000 key words and place names from the text.

• *Annotated Web Links*

• *Area and Demographic Data*

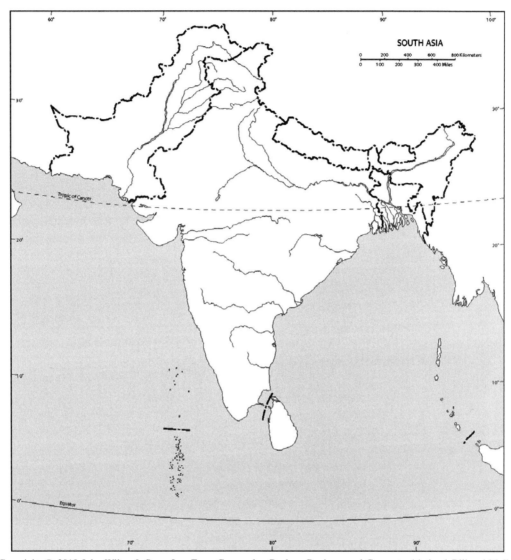

SOUTH ASIA

0   200   400   600   800 Kilometers
0   100   200   300   400 Miles

Tropic of Cancer

Equator

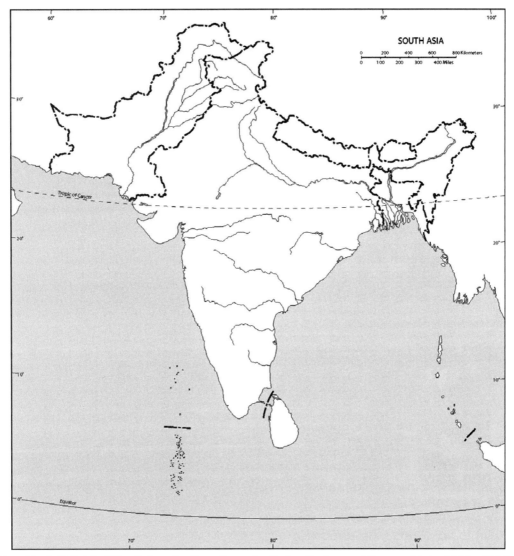

SOUTH ASIA

0   200   400   600   800 Kilometers
0   100   200   300   400 Miles

Tropic of Cancer

Equator

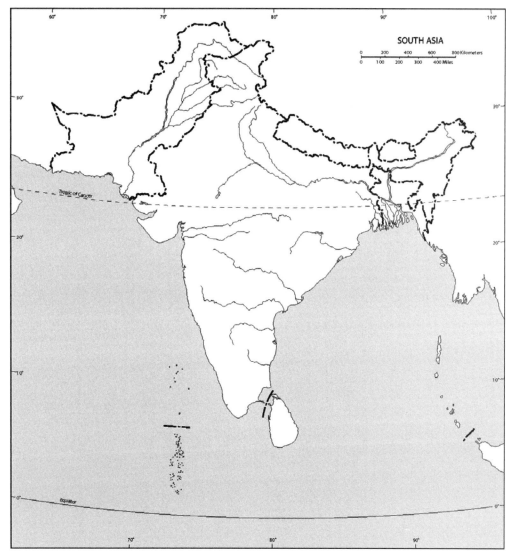

SOUTH ASIA

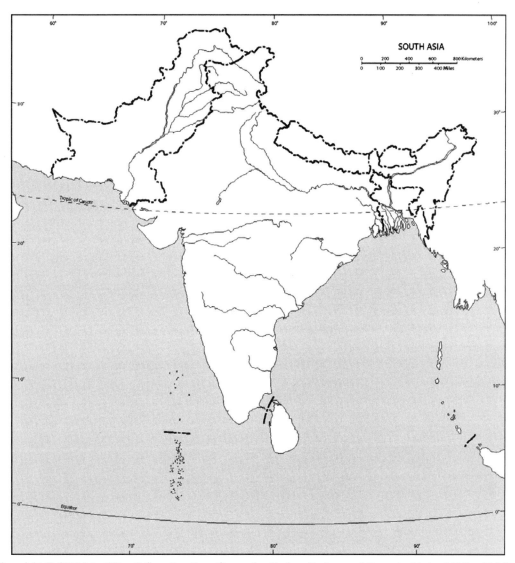

SOUTH ASIA

0   200   400   600   800 Kilometers
0 100   200   300   400 Miles

Tropic of Cancer

Equator

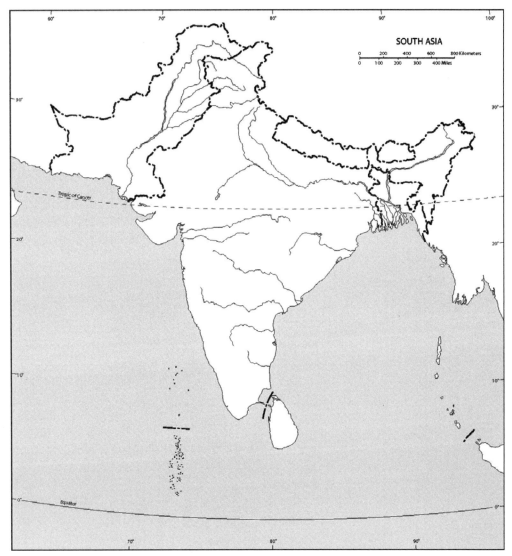

SOUTH ASIA

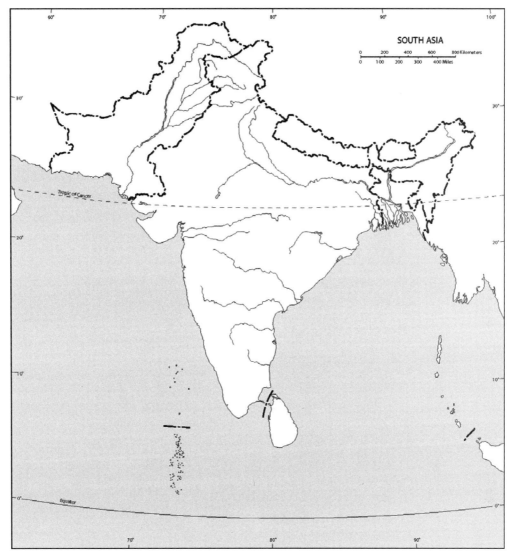

SOUTH ASIA

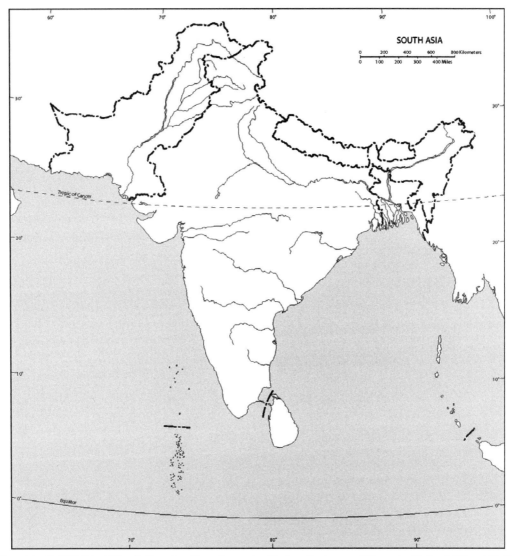

SOUTH ASIA

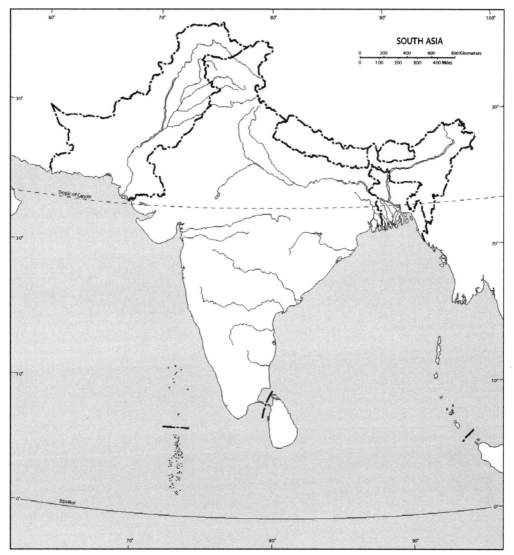

SOUTH ASIA

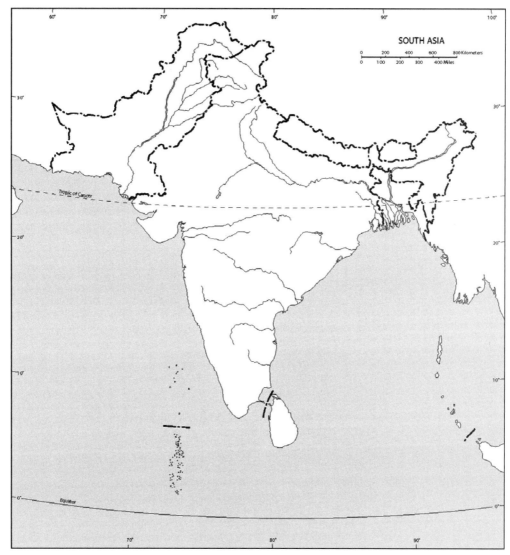

SOUTH ASIA

Tropic of Cancer

Equator

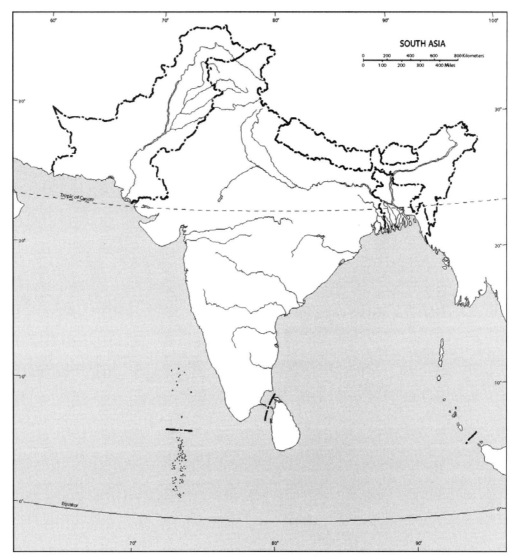

SOUTH ASIA

Tropic of Cancer

Equator

CHAPTER #9   EAST ASIA

SECTION 9.1: Map Creation

In this section you will create your own detailed maps of the different major physical, cultural, geopolitical and environmental attributes of the realm. Each of these maps is critically important for understanding the Major Geographic Qualities of East Asia, including their spatial distribution and functionality. By creating these on your own using the blank maps provided at the end of this chapter, you will actively learn where these different features are located and how their spatial distribution affects the ways of life in each of these respective areas. Blank Maps for East Asia are provided for you in Section 9.3. Label each of your maps with the appropriate name and title. For instance, the first map you will create will be called 'Map 9–1: Countries of East Asia'.

Map 9–1: Countries of East Asia

On the map, label the following countries of the East Asian Realm. To assist with creation of this map, refer to pp. 454–455.

- China
- North Korea

- Japan
- South Korea

- Taiwan

- Mongolia

Map 9–2: Major Cities of East Asia

On this map, label the following major cities of the East Asian Realm. You may need to use abbreviations and/or arrows if you do not have enough space to write in the full name on the map. To assist with creation of this map, refer to the overall map of East Asia found on pp. 454–455.

- Beijing
- Wuhan
- Lhasa
- Hiroshima
- Pyongyang

- Shanghai
- Zhengzhou
- Tokyo
- Nagasaki
- Ulaanbaatar

- Hong Kong
- Changchun
- Kobe
- Taipei

- Guangzhou
- Xianyang
- Fukuoka
- Seoul

Map 9–3: Physical Features of East Asia

East Asia has a wide diversity of physical features throughout the realm that have played an important role in settlement patterns and political boundaries for thousands of years. For this map, you will draw in the physical features listed below. Use a different colored pencil or pen for each category, and to assist with creation of this map, refer to both pp. 454–455 and Figure 9–3, p. 460.

Mountain Ranges:
- Sikhote-Alan Range
- Wuyi Shan
- Daxue Mountains

- Greater Khingan Range
- Altai Mountains
- Himalayas Mountains

- Luliang Shan
- Kunlun Mountains
- Tian Shan Mountains

Plateaus and Basins:
- Tarim Basin
- Mongolian Plateau
- Northeast China Plain

- Qinghai-Xizang (Tibet) Plateau
- Yunnan Plateau
- North China Plain

- Junggar Basin
- Sichuan Basin
-

Deserts:
- Gobi Desert
- Takla Makan Desert
- Ordos Desert

Peninsulas and Islands:
- Ryukyu Islands
- Kurile Islands
- Hainan Island
- Hokkaido Island
- Honshu Island
- Kyushu Island
- Ryushu Island
-

Map 9–4:   *East Asian Bodies of Water*

East Asia is a realm that has been influenced by its water resources, both in terms of spatial history and also economic development. On this map, you will identify the major bodies of water found inside this realm. These are divided into groups and it is recommended to use a different colored pencil or pen for each of the names per water body category. Use pp. 454–455 as a guide.

Major Rivers and Lakes:
- Salween River
- Mekong River
- Chang Jian River
- Hongshui River
- Huang He River
- Brahmaputra River
- Chang Jiang (Yangtze) River
- Three Gorges Dam
- Yalu River

Major Saltwater Bodies:
- East China Sea
- Yellow Sea
- Sea of Japan
- Philippine Sea
- Strait of Taiwan
- Pacific Ocean
- Bo Hai Gulf
- Korea Bay
- South China Sea
- Luzon Strait
- Tsugaru Strait
- Gulf of Tonkin

Map 9–5:   *Major Climate Regions of East Asia*

East Asia is a realm that contains a wide variety of climates, changing dramatically from east (Mild and Cold Mid-Latitude Climates) to west (Dry and High Climates). Use Figure 9–7 (p. 469), Climates of China, and Figure G – 7 (pp. 16–17) as your guide. The major Köppen Climate types are given below for each of the climates found in this realm. Using a different colored pen/pencil for each category, outline and then color in these climate types:

Tropical Climates:
- Aw
- Am

Dry Climates:
- BWk
- BSk

Mild Mid-Latitude Climates:
- Cfa
- Cwa
- Cwb

Cold Mid-Latitude Climates:
- Dwa
- Dwb
- Dfb

High Climates:
- H

Map 9–6:   *East Asian Population Densities*

The human populations across the East Asian Realm are very unevenly distributed and range from some of the most dense to the most sparse in the world. Using the East Asian Population Distribution map on

Figure 9–2 (p. 458), we see that five general population density categories exist in the East Asian Realm. Using a different colored pen/pencil for each category, outline and color the countries/sections of countries that fit into each of these designations.

Very Dense: The highest density populations are found along the eastern coastline of China (to an area about 200 miles inland), and also occupy all of Japan, South Korea and Taiwan. Over 1 billion people occupy this territory.

Dense: This population category extends inland from the Very Dense populations in China to about the middle of the country.

Sparse: This category extends from central China to West-Central China, to the region of the western basins and plateaus.

Very Sparse: Found throughout Mongolia, and also northern, western and southwestern China.

Empty: Empty populations are found in Northwestern China and Western Mongolia, where there is almost no permanent population due to the harsh climatic conditions.

Map 9–7: *Spatial History: The Evolution of the Chinese Empire*

Using Figure 9–4 (p. 463), you will create a map that shows the spatial development of the Chinese Empire from 2,000 years ago through present-day. Once completed, you will see the spatial trend of conquering through time that expanded the land area of Chinese domain across the continent of Asia.

Category 1:   *Earliest Core Area (settled by 1900 B.C., also known as the Xia Empire)*

Category 2:   *Shang Dynasty (1766–1080 B.C.), added land to the regions around the Xia Empire*

Category 3:   *Han Dynasty (200 B.C. – 220 A.D.), added significant territory to China by spreading west and south across the region.*

Category 4:   *The Great Wall of China (built by the Han Dynasty); draw in where this wall was built.*

Category 5:   *Qing Dynasty (1644–1911 A.D.), resulted in tremendous land area conquering that spills across China's present-day borders.*

Map 9–8: *Ethnolinguistic Regions of East Asia*

Using Figure 9–11 (p. 480), you will create a map that shows the spatial distribution of languages and associated cultures throughout East Asia. Use a different colored pen/pencil for each language represented, and notice the spatial complexity emerge across Asia from west to east and the overall dominance of the Sino – Tibetan Family.

Sino – Tibetan Language Family:

- Northern Mandarin	- Southern Mandarin	- Tibetan
- Wu	- North Min	- South Min
- Hakka	- Yue (Cantonese)	- Xiang
- Gan	- Miao – Yao	- Thai
- Korean	- Manchu	

Altaic Language Family:
- Mongolian   - Turkic

Austro – Asiatic:
- Mon – Khmer

Indo – European:
- Tajik

Map 9–9:   Agricultural Regions in East Asia

Using Figure 9–14 (p. 487) as a guide, circle and label the following areas on your map, displaying where these major economic activities are found today. Use a different color pen/pencil (if possible) for each area, and note the spatial difference in agricultural locations as you travel from east to west across the realm (due to changes in climate conditions).

- Rice and Wheat	- Rice and Tea	- Double Rice Crop
- Rice in Sichuan	- Upland Rice	- Winter Wheat and Millet
- Spring Wheat	- Xizang Pasture	- Winter Wheat and Gaoliang
- Inner Mongolian Pasture	- Oasis Farming	- Spring Wheat and Soybeans

Map 9–10:   Economic Resources in East Asia

China is emerging as a world leader in both coal production and consumption, and their ever-increasing economy with its exploding manufacturing empire requires more power every day. Massive coal reserves are found in China at different locations, along with oil and natural gas deposits. On this map you will show where these energy sources are located using a different colored pen/pencil for each type. You will also detail where the primary locations of manufacturing are found by circling the general regions of their presence. Refer to Figure 9–15 (p. 489) and the descriptions below as your guide.

Coalfield Regions: There are many established coalfields in the country, as seen in the textbook:
- South-Central China, near the city of Chongqing.
- North-Central China, centered around Taiyuan.
- Northwestern China, near the city of Urumqi.

Gas Field Regions: There are two major clusters of gas fields:
- East – Central China, near the cities of Chengdu and Zhengzhou.
- North China, centered on the city of Yumen.

Oilfield Regions: Oilfields are found in the following four major locations:
- Northwestern China, near the city of Urumqi.
- Northeastern China, near the city of Harbin.
- Central China, from Chengdu to Taiyuan.
- Southeastern China, both onshore and offshore of Hong Kong.

Manufacturing Regions: Three major clusters of manufacturing exist in the country:
- Northeastern China, from Beijing to Shenyang.
- Eastern China, from Shanghai to Wuhan.
- Southeastern China, centered around Hong Kong.

Map 9–11:   Environmental Issues in East Asia

The East Asian Realm is one of the most environmentally-degraded places in the world today. This realm is one of the most advanced realms in the world today in terms of technology and emissions standards for air and water quality. However, moderate to severe environmental issues persist in certain locations.

These environmental issues are divided into categories below. Using a different colored pen/pencil for each, circle or highlight the areas and label them using their respective environmental situation.

Sealevel Rise: The lowland regions from Shanghai to Hong Kong in China are the most susceptible to sealevel rise, as these areas are only 5–10 feet in elevation near the coastline. A rise of 10–20 feet in sea level over the next 100 years could displace up to 50 million people and cause tremendous loss of agricultural lands as well.

Coastal Pollution: Almost all of the East China and South China Seas are affected by coastal pollution, sources primarily being urban sewage and manufacturing waste that are discharged directly into this body of water due to lax environmental regulations.

River Pollution: Almost all of East Asia's major rivers have water quality issues ranging from moderate to excessively toxic. In particular, the Yangtze, Huang He and Mekong Rivers all are considered severely contaminated near the coastline. Trace these using a blue pen/pencil and label them accordingly.

Air Quality Issues: As a result of the heavy manufacturing and industry presence in the East Asian Realm along with increasing urban population concentrations, poor urban air quality is a severe problem across the realm. Draw a circle around the worst areas of air quality, enclosing the eastern coastline of China and South Korea, all of Japan, and all of Taiwan. Additionally, draw a second circle around the north-central locations in China, where you will find substantial coal mining and oil refining industries.

Map 9–11:   Regions of the East Asian Realm

Using Figure 9–5 (p. 467) in the textbook, color the following regions of the East Asian Realm and use a different colored pen/pencil for each region and the country list below as a guide. After completing this exercise, consult the textbook to understand why each of these is considered a separate region of the realm.

- China Proper   - Xijiang (Tibet)   - Xinjiang
- Mongolia   - Jakota Triangle

SECTION 9.2: Review Questions

In this section you will review the main concepts and terminology from the East Asian Realm by answering the questions provided below. Write in your answers in the space provided, and refer to the respective sections/pages of the chapter for the correct information.

Review Section 9.2.1 *East Asia's Major Geographic Qualities (p. 457)*

Summarize the eleven Major Geographic Qualities of East Asia in your own words:

  1.)

  2.)

  3.)

  4.)

  5.)

  6.)

  7.)

  8.)

  9.)

  10.)

  11.)

Review Section 9.2.2 *Defining the Realm (p. 457)*

1.) In what ways is East Asia currently experiencing a 'massive geographic transformation', as stated in the textbook?

2.) What six primary countries constitute this realm?

3.) How is China different from Japan today in terms of physical area and population? How did this difference vary throughout the 1900's?

Review Section 9.2.3 *Natural Environments (pp. 459–461)*

1.) How does China vary physically from east to west? What is the resulting geographic impact on the population of this country?

2.) Building on Question #1, the physical regions are not the only factor limiting or promoting population distribution in the realm. How do the climates vary from east to west, and why?

3.) What are the Three Great Rivers of the realm, and where are they located?

4.) How is the physiography of Japan, The Koreas and Taiwan all similar? In what way(s) does this vary from China and Mongolia?

Review Section 9.2.4   East Asia's Historical Geography (pp. 461–465)

1.) When were the first agricultural systems in this realm, and what primary crops were involved?

2.) What is the process of 'state formation', and how does it relate to China's historical dynasties?

3.) What were the ten main Chinese Dynasties, starting with the Xia Dynasty? When were these in existence? List and give brief details about each.

A.)

B.)

C.)

D.)

E.)

F.)

G.)

H.)

I.)

J.)

4.) Summarize the 'Points to Ponder' in your own words, and why you think each will be important geopolitical and/or environmental issues in the 21st century.

A.)

B.)

C.)

D.)

Review Section 9.2.5   Regions of the Realm (p. 466)

1.) Name the five regions of the realm, and the countries within each:

A.)

B.)

C.)

D.)

E.)

2.) What are the four largest cities in this realm in terms of population? Name these and the respective countries that these belong to.

Review Section 9.2.6   China Proper: Location, Extent and Environment (pp. 466–469)

1.) Compare China's Empire to that of the European Empires in the 13th century. How did China view the Europeans in comparison to their own civilization?

2.) When did China become open to visitors and tourists? What was the role of isolation in China's affairs from historic times to present-day?

3.) What role has Japan played in the economic and technological success of the realm in recent times?

4.) Compare China to the United States in terms of physical size and climate variation. How are the two countries similar, yet very different?

Review Section 9.2.7   Evolving China (pp. 469–475)

1.) How long has the Chinese culture been in existence, according to the textbook? What is sinicization, and how does it relate to this idea?

2.) Who was Confucius, when did he live and what impact did ha have upon the Chinese society?

3.) What was the attitude of the Chinese people toward the European arrival in the 1300's, especially in terms of the goods the Europeans had to offer?

4.) What impact did the fall of the Manchu (Qing) Dynasty have upon trade and political ties with the European nations? When did this occur, and how was the drug Opium involved in this event?

5.) What countries claimed land area in China during the late 1800's, and where were these located?

6.) Describe the evolution of China geographically from the late 1800's until the start of World War II, by decade. Who were the major rulers of the country during this time, and what impact did they have upon the realm spatially and politically?

A.) 1890–1899:

B.) 1900–1909

C.) 1910–1919

D.) 1920–1929

E.) 1930–1939

7.) What does 'Extraterritoriality' mean, and how did it affect the geography of China?

8.) Describe China after World War II and its political shift from democratic to communism. What ruler proclaimed the start of the 'People's Republic of China' in 1949?

Review Section 9.2.8   China's Human Geography (pp. 475–479)

1.) Describe the pros and cons of the communist regime in China, in terms of the progress and setbacks this government offered to the country after 1949.

2.) What are the events knows as 'The Great Leap Forward' and 'The Great Proletarian Cultural Revolution'? How did both impact China's development politically and socially?

3.) Name the four general political and administrative units of China, and briefly describe each:

A.)

B.)

C.)

D.)

4.) What is the estimated population of China today? Where does this rank in terms of world populations, and how many people per year are being added to this population?

5.) What was the 'One Child' policy of China, when was it instituted and for what reasons? Was it successful in curbing China's explosive population growth?

6.) What is China's current birth rate today, and does it favor one particular gender over another? What factors are involved in this gender bias?

Review Section 9.2.9   People and Places of China Proper (pp. 479–490)

1.) What spatial relationship is seen in China's population pattern?

2.) Which four river basins are the center of Chinese population density? What percentage of the population of the country do these contain?

3.) Compare and contrast the Northeast China Plain with the North China Plain in terms of location, physical features, population density and cultural diversity.

4.) Where are the cities of Tianjin and Beijing respectively located, and what role do they play in the current cultural and economic geography of China?

5.) Where is 'Inner Mongolia', and how is it different from the previously-discussed subregions of China?

6.) Describe the Yangzi (Yellow) River in terms of location and its impact upon the populations of China agriculturally and economically. Include the Three Gorges Dam as part of your discussion.

7.) Where is the Sichuan Basin, and why is it economically important for China today?

8.) Compare Hong Kong to Shanghai in terms of location and the respective rivers these are situated upon. How are these great cities similar, yet different culturally and politically?

Review Section 9.2.10   Xizang and Xijiang (pp. 492–493)

1.) Compare Xizang to Xijiang in terms of spatial location, physical features, remoteness from the most populated regions of China, and current geopolitical issues:

A.) Xizang:

B.) Xinjiang:

Review Section 9.2.11   China: Global, Local, and Superpower (pp. 493–502)

1.) How has the development of modern China influenced the concept of the Pacific Rim? Give examples of the improvements in technology and advancement throughout the country to support your answer.

2.) What are the two major consequences of this development in China? Do these have any sort of spatial component, and if so, where?

3.) What is the 'Geography of Development'? What measures are used to determine achievement in a country along these lines? Give at least five examples, as seen in the textbook.

4.) What is 'Neoliberalism', and how is it affecting the economic status of China today?

5.) Define the term 'Special Economic Zones', and the purpose of their creation.

6.) Where are China's primary seven SEZ's located today? Name these and briefly describe each:

A.)

B.)

C.)

D.)

E.)

F.)

G.)

7.) Describe the role of Hong Kong in China's economy today, as both a financial leader and economic giant. When did the transformation of this former fishing village to world city take place?

8.) With all of the economic surging of the Chinese economy, what potential trouble spots may be on the horizon for this country, from economic to political issues?

9.) How strong is China in terms of its political and military power today? Give specific examples.

10.) Explain what is meant by the idea that China could be the site of the first 'Intercultural War' of the 21st century.

11.) How has the concept of nationalism helped to continue the modernization of China today?

Review Section 9.2.12 Regions: Mongolia and Taiwan (pp. 502–505)

1.) Compare Mongolia to China in terms of physical size and population. How different are these two countries?

2.) What involvement has Mongolia had with China and the Soviet Union in the 1900's? What effect did the fall of the USSR have upon Mongolia's political structure?

3.) What are the main economic pursuits in Mongolia? Is it primarily agricultural-based, or are there other activities within this region?

4.) What is the present relationship between China and Taiwan? How did this situation happen over time?

5.) What is the physical size, location and population of Taiwan today?

6.) In your opinion, should Taiwan be another S.A.R. of China, or should it remain independent? Why?

7.) Describe the role of Taiwan today in the global economy, using the term 'Economic Tiger' in your discussion. How did it get to this point?

Review Section 9.2.13 Japan (pp. 505–520)

1.) Describe Japan's economic recovery from World War II to present-day. What geographic factors allowed Japan to rebuild into the world superpower it is today?

2.) What was the Meiji Restoration, and how did it allow Japan to develop in the 1800's? What role did Japan's relative location have upon this process?

3.) Japan is very different physically than any other country in the realm. Describe some of these physical differences, both in terms of land area, features, and climate.

4.) What is the estimated population of Japan today? What percentage of the land area of Japan is habitable? Why would this be a problem?

5.) Define the term 'Areal Functional Organization', and the role it played in Japan's regional organization.

6.) Name and briefly describe the four main economic regions of Japan, in terms of where they are located, major cities, and economic importance.

A.)

B.)

C.)

D.)

7.) Summarize Tokyo in terms of location, population, culture, economic and financial importance.

8.) Describe some of the current geopolitical issue in Japan today, focusing on the ideas of limited land area and agricultural concerns. What steps are being taken to remedy these issues, if anything?

9.) What does Japan's infrastructure need in the 21st century, in order to have an economic turnaround?

10.) Are the populations of Japan expected in increase, stay the same, or decrease in the 21st century? Give specific examples of projections in the textbook. What effect will this have upon the Japanese economy?

Review Section 9.2.14   The Koreas (pp. 520–525)

1.) Describe the circumstances leading to the creation of North and South Korea after World War II, and the immediate differences between these two countries politically and economically.

2.) Summarize the Korean War: When and why did it happen, and what have been the results spatially and upon the people of these two countries?

3.) Compare South Korea to North Korea in terms of land area and population. How might the idea of 'regional complementarity' have helped these two countries, if not for the political divisions?

4.) What events led South Korea to become a world economic power from the end of World War II through present-day? Use the term 'State Capitalism' in your explanation.

5.) What are some of the main products manufactured in South Korea? Name the three main industrial regions and summarize each in terms of goods and services.

A.)

B.)

C.)

Review Section 9.2.15   A Jakota Triangle? (pp. 525–527)

1.) Where is the 'Jakota Triangle', and what factors may prompt its creation as a region of East Asia?

2.) Name and describe at least three challenges this area faces in the 21st century. Why is this part of East Asia not presently considered a region of the realm today?

SECTION 9.3:   Blank Maps of East Asia

Blank maps are provided for you to utilize with the mapping questions from Section 9.1 of the Study Guide. You are being provided with one blank map for each mapping exercise given in Section 9.1. It is a good idea to make additional copies of these blank maps to use for extra practice and review.

SECTION 9.4:   Student Companion Website

Additional study tools are available on the Student Companion Website at www.wiley.com/college/deblij. Features include:

- *Flashcards* offer an excellent way to practice key concepts, ideas, and terms from the text. You can review and quiz yourself on the concepts, ideas, and terms discussed in each chapter.

- *Map Quizzes* help you to master the place names for the various regions studies. Three game-formatted map activities are provided for each chapter.

- *Chapter Review Quizzes* provide immediate feedback to true/false, multiple choice, and short answer questions.

- *Audio Pronunciation* is provided for over 2000 key words and place names from the text.

- *Annotated Web Links*

- *Area and Demographic Data*

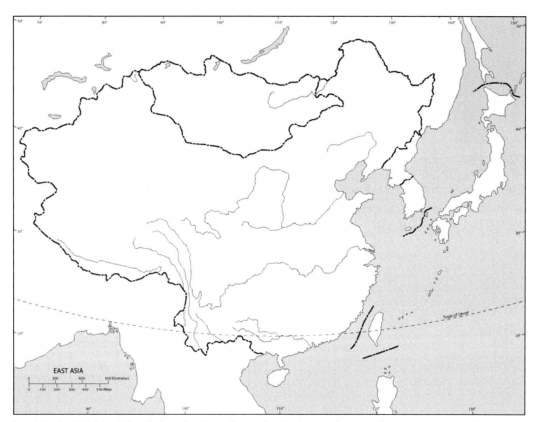

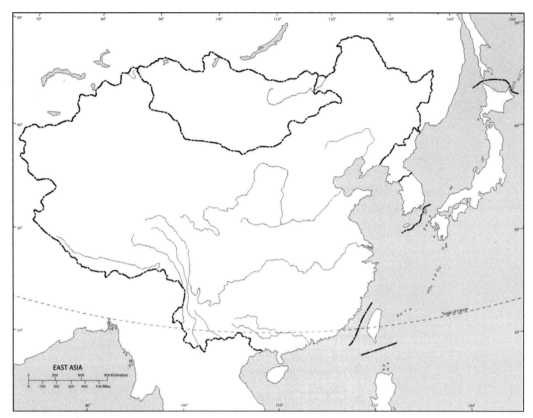

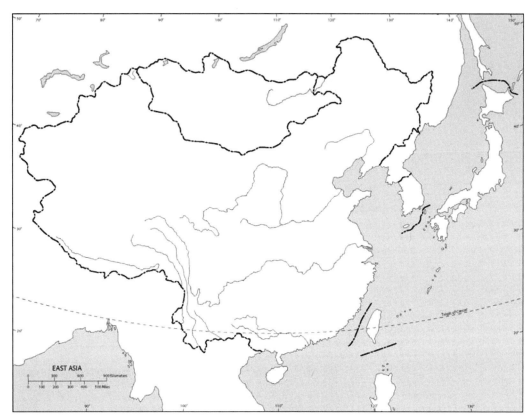

EAST ASIA

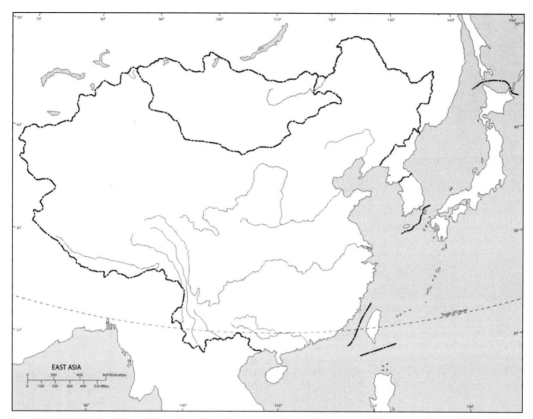

EAST ASIA

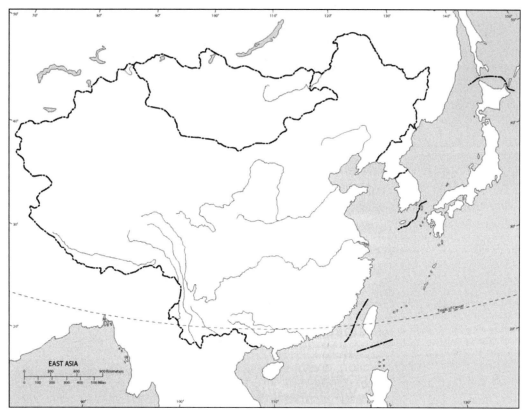

EAST ASIA

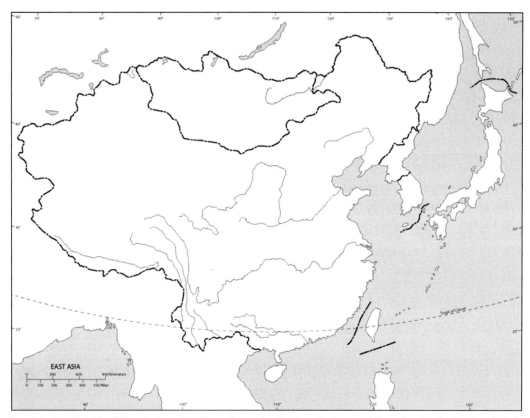

EAST ASIA

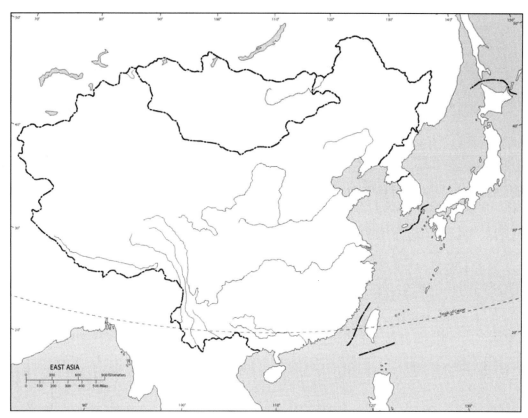

EAST ASIA

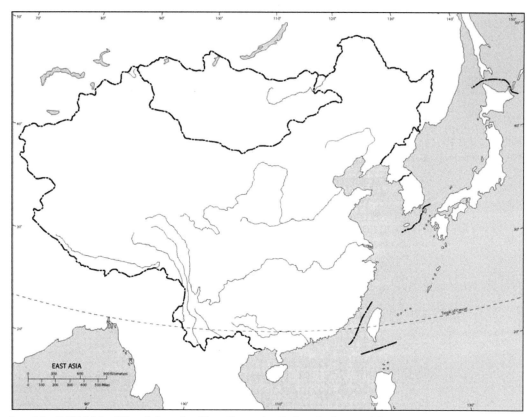

EAST ASIA

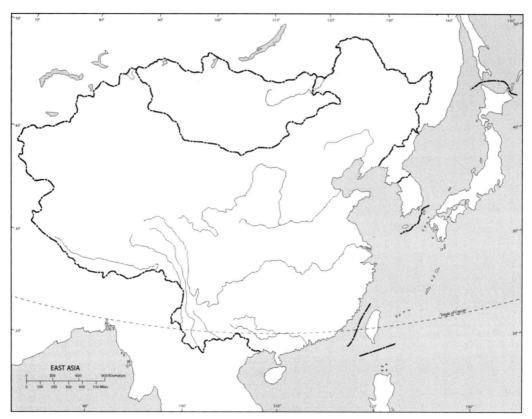

CHAPTER #10   SOUTHEAST ASIA

SECTION 10.1:   Map Creation

In this section you will create your own detailed maps of the different major physical, cultural, geopolitical and environmental attributes of the realm. Each of these maps is critically important for understanding the Major Geographic Qualities of Southeast Asia, including their spatial distribution and functionality. By creating these on your own using the blank maps provided at the end of this chapter, you will actively learn where these different features are located and how their spatial distribution affects the ways of life in each of these respective areas. Blank Maps for Southeast Asia are provided for you in Section 10.3. Label each of your maps with the appropriate name and title. For instance, the first map you will create will be called 'Map 10–1: Countries of Southeast Asia'.

Map 10–1:   Countries of Southeast Asia

On the map, label the following countries of the Southeast Asian Realm. These have been divided into two regions, and you may need to use abbreviations and/or arrows if you do not have enough space to write in the full name on the map. To assist with creation of this map, refer to pp. 528–529.

Countries in Mainland Southeast Asia:
- Vietnam   - Cambodia   - Laos
- Myanmar   - Thailand   - Singapore

Countries in Island Southeast Asia:
- Indonesia   - Malaysia   - Brunei
- East Timor   - The Philippines

Map 10–2:   Major Cities of Southeast Asia

On this map, label the following major cities of the Southeast Asian Realm. You may need to use abbreviations and/or arrows if you do not have enough space to write in the full name on the map. To assist with creation of this map, refer to the overall map of Southeast Asia found on pp. 528–529. Write in the city name using a different colored pencil or pen for each of the regions these belong.

Cities in Mainland Southeast Asia:
- Hanoi   - Saigon (Ho Chi Minh City)   - Nha Trang   - Da Nang
- Phnom Penh   - Viangchan   - Bangkok   - Yangon
- Phuket   - Kuala Lumpur   - Singapore

Cities in Island Southeast Asia:
- Jakarta   - Manila   - Medan   - Palembang
- Surabaya   - Dili   - Bandar Seri   - Kuching
- Davao   - Ujungpandang   - Manado

Map 10–3:   *Physical Features of Southeast Asia*

Southeast Asia contains major physical features that have defined settlement, agriculture and population distribution throughout time. For this map, you will draw in the physical features listed below, using a different colored pencil or pen for each of the names per category. To assist with creation of this map, refer to both the overall map of Southeast Asia found on pp. 528–529, and also Figure 10–3, p. 534.

Mountain Ranges: Using a blue pen/pencil, draw in the locations of these mountain ranges on your map and label them accordingly. Use a Λ symbol to denote the locations of these features.

- Arakan Mountains   - Annam Mountains   - Bago Mountains
- Barisan Mountains   - Maoke Mountains   - Molucca Mountains

Islands: There are hundreds, if not thousands, of islands in this realm. Locate the major islands below, either writing the name on them or circling the island and labeling it on the side of your map.

- Sumatra   - Java   - Bali
- Timor   - Lombok   - Sumbawa
- Flores   - Sumba   - Sulawesi
- Moluku Islands   - New Guinea   - Borneo
- Mindanao   - Spratly Islands   - Samar
- Panay   - Mindoro   - Luzon
- Nicobar Islands   - Andaman Islands   - Halmahera

General Physical Features:

Khorat Plateau   - Shan Plateau   - Malay Peninsula
Tonkin Plain   - Kra Isthmus   - Kalimantan Plateau

Map 10–4:   *Southeast Asian Bodies of Water*

Southeast Asia is a realm that has been influenced by its water resources, both in terms of spatial history and also economic development. On this map, you will identify the major bodies of water found inside this realm. These are divided into groups and it is recommended to use a different colored pencil or pen for each of the names per water body category. Use pp. 528–529 as a guide.

Major Rivers in the Realm:

- Salween River   - Mekong River   - Irrawaddy River
- Chao Phraya River - Red River

Major Saltwater Bodies near Mainland Southeast Asia:

- Gulf of Thailand   - Andaman Sea   - Indian Ocean
- Gulf of Martaban   - Bay of Bengal   - Gulf of Tonkin
- South China Sea   - Strait of Malacca

Major Saltwater Bodies near Island Southeast Asia:

- Sulu Sea   - Philippine Sea   - Celebes Sea
- Maluku Sea   - Banda Sea   - Arafura Sea
- Java Sea   - Bali Sea   - Flores Sea
- Timor Sea   - Suva Sea   - Sunda Strait

Map 10–5:   Major Climate Regions of Southeast Asia

Southeast Asia is a realm that contains a very homogeneous climate distribution, with tropical climates dominating the realm due to its situation on/near the equator. High Climates are found in some island locations, and Mild Mid-Latitude Climates are seen in the extreme northern regions of the realm. Use the World Climate Map (Figure G – 8, Introductory Section) as your guide. The major Köppen Climate types are given below for each of the climates found in this realm. Using a different colored pen/pencil for each category, outline and then color in these climate types:

Tropical Climates: *Mild Mid - Latitude Climates*:
- Am   - Cwa
- Af
- Aw

High Climates:
- H

Map 10–6:   Southeast Asian Population Densities

Using the Southeast Asian Population Distribution map on Figure 10–4 (p. 535), we see that three general population density categories exist in the Southeast Asian Realm. Using a different colored pen/pencil for each category, outline and color the countries/sections of countries that fit into each of these designations. Once completed, you will see a spatial pattern of population distribution emerge across the realm, with the heaviest populations in the island regions of Indonesia and The Philippines.

Very Dense: The heaviest population concentrations are found in the following locations:
- The Philippines
- The Islands of Sumatra, Java, Bali and the Lesser Sunda Islands of Indonesia
- North Vietnam
- South Thailand, centered on Bangkok

Dense: This population category accounts for the following:
- The rest of Mainland Southeast Asia except for the northern regions, and
- The rest of Island Southeast Asia, except for the island of Borneo

Sparse: The only sparse populations in the realm are found in two locations:
- The center of the Island of Borneo
- Northern Myanmar and Laos

Map 10–7:   Spatial History: Colonial Influences in Southeast Asia

Using Figure 10–7 (p. 539) and the descriptions below, you will create a map that shows the colonial history and development of Southeast Asia. Use a different colored pen/pencil for each colonial influence in the realm.

British:	Settled and occupied the present-day countries of Myanmar and Malaysia, until pulling out of these areas in 1947.
French:	Colonized the present-day countries of Vietnam, Laos and Cambodia, and remained in the region until the 1950's.
The Netherlands:	Responsible for settling all of Indonesia in the 1500's, and relinquished control in 1946 after World War II.
Spanish:	Settled The Philippines and controlled this region until 1898.
The United States:	Assumed control over The Philippines from 1898 until 1946 and the end of World War II, when the country gained its full independence.

Map 10–8:   Religions in Southeast Asia

Every major religion in the world, with the exception of Judaism, is represented in Southeast Asia. From significant Buddhist and Christian populations, to the highest density of Muslims in the world, the religious diversity of Southeast Asia and its relation to past and present geopolitical issues cannot be understated. You will create a map that shows the religious diversity of Southeast Asia, using the descriptions below.

Buddhists:	Located on the mainland in the countries of Vietnam, Cambodia, Thailand, Myanmar (except for the northern regions) and in southern Laos.
Muslims:	The greatest density of Muslims in the world, in terms of people per square mile, is found in this realm in both Indonesia and Malaysia.
Hindus:	Found on the Island of Bali in Indonesia in great concentrations.
Catholics:	Primarily located in the Philippines, as a result of Spanish settlement.
Christian (Protestant):	The main sect of Protestant Christianity is found in the middle of the Island of Sumatra, in Indonesia.
Tribal Religions:	These are found in two locations: • Northern mainland Southeast Asia, primarily in the northern regions of Laos and Myanmar. • The Island of Borneo, primarily in the highland central areas.

Map 10–9:   The Ethnic Mosaic of Southeast Asia

Using Figure 10–5 (p. 536) as a guide, you will create a map showing the ethnic distribution of people across the Southeast Asian Realm. Use a different colored pen/pencil for each group, and notice the spatial pattern of ethnic groups as you progress through the map- why do you think these groups chose the regions you see for settlement and development? Refer to the textbook for these answers and additional information about the topic.

Indo – Aryan: Found in a small region of mainland Malaysia, near Singapore.

Chinese: Located primarily on the western coastline of mainland Malaysia, and also the northern coastline of the Island of Borneo (Malaysia and Brunei).

Thai: Found throughout almost all of Thailand, into Eastern Myanmar, Northern Laos and Northern Vietnam.

Miao – Yao: Located mainly in Northern Vietnam, along the China border.

Tibeto – Burman: Located throughout the central and western regions of Myanmar.

Mon – Khmer: Found in most of Cambodia, extending into southern Vietnam and southern Laos.

Vietnamese: Located throughout the rest of Vietnam, excluding those regions occupied by the Thai and Mon – Khmer ethnic groups.

Papuan: A small concentration of this ethnic group is found in extreme Eastern Indonesia, and extends into the Island of New Guinea.

Indonesian: The largest ethnic group in the realm in terms of population and spatial coverage, this concentration is found throughout the rest of Indonesia, Malaysia and the Philippines.

Map 10–10:   Environmental Issues in Southeast Asia

Southeast Asia has seen significant environmental degradation over the past 50 years as a result of globalization and the increase in manufacturing jobs throughout the realm. In this map you will circle or highlight the regions affected by the following environmental situations using a different colored pen/pencil for each, and the descriptions below as a guide.

Sealevel Rise: The low islands of Indonesia (eastern region) are the most susceptible to sealevel rise, as many of these are only 5–15 feet above sea level.

Coastal Pollution: There are four main locations of significant coastal pollution found in the realm:
• Coastlines of Northern Vietnam, centered near Hanoi.
• Southern Vietnam to Bangkok, in the Gulf of Thailand.
• The coastal waters off the islands of Sumatra, Java and Bali.
• The coastal waters off Manila (The Philippines).

River Pollution: Due to increased use of fertilizer and pesticides in agriculture along with very poor environmental standards for dumping of manufacturing waste, all of the major rivers on the Mainland of Southeast Asia have moderate to severe river pollution. Trace these using a blue pen/pencil and label them accordingly.

Geologic Hazards: The 2004 Tsunami that caused nearly 300,000 casualties was generated just offshore of the northwestern coast of the Island of Sumatra. As it turns out, the entire island regions of Indonesia, Malaysia and The Philippines are the most geologically-active in the world. It is very common to for these offshore regions to experience a 6.0 earthquake every week (if not every day). Additionally, the island region is composed almost entirely of active volcanoes (the eruption of Mt. Pinatubo in 1991 being the most recent and prominent example).

SECTION 10.2: Review Questions

In this section you will review the main concepts and terminology from the Southeast Asian Realm by answering the questions provided below. Write in your answers in the space provided, and refer to the respective sections/pages of the chapter for the correct information.

Review Section 10.2.1   Southeast Asia's Major Geographic Qualities (p. 531)

Summarize the nine Major Geographic Qualities of Southeast Asia in your own words:
  1.)
  2.)
  3.)
  4.)
  5.)
  6.)
  7.)
  8.)
  9.)
 10.)
 11.)

Review Section 10.2.2   Defining the Realm and the Regional Framework (pp. 531–535)

1.) Summarize the physical setting of Southeast Asia, and the ways in which it could be compared to Eastern Europe.

2.) Southeast Asia is divided into two primary regions. Name these two regions and the countries that comprise each:

   A.)

   B.)

3.) Though there is not currently a dominant country in this realm, which country has the highest population and greatest amount of land area?

4.) What impact did the Europeans have upon unifying the political and cultural geography of the realm? Give an example of a country that has been affected by this intervention.

Review Section 10.2.3   Southeast Asia's Physical Geography (pp. 535–537)

1.) Name the major mountain chains found in this realm, and describe their respective location. Are volcanic mountains found in both the mainland and island regions?

2.) What are the four main rivers found in Southeast Asia? Name these and briefly describe their location.

Review Section 10.2.4   Population Geography (pp. 535–537)

1.) Are the populations of Southeast Asia evenly or unevenly distributed? Give examples from the textbook to support your answer. Which two countries have the largest populations?

2.) Name and briefly describe the nine ethnic groups of Southeast Asia:

A.)

B.)

C.)

D.)

E.)

F.)

G.)

H.)

I.)

3.) Immigrants have come to this realm primarily from which country? Why did the Europeans encourage this movement, and what effect did this minority have upon cities in the realm?

Review Section 10.2.5   How the Political Map Evolved (pp. 537–543)

1.) Name the four main colonial powers that settled and occupied this realm, and describe where each was located. When did each relinquish control of these areas?

A.)

B.)

C.)

D.)

2.) Where is 'Indochina' located, and why is this term appropriate for these areas in Southeast Asia?

3.) Describe the religious in Southeast Asia: What are the main religions found here, and where are they primarily located today?

4.) Summarize the Chinese presence in Southeast Asia during the time of European colonization. After the European countries withdrew, what was the impact upon the Chinese populations here?

5.) Read the debate on the Chinese Presence in Southeast Asia (p. 542), and give your opinion on the situation: Are the Chinese indispensable or too influential in the realm today, or both?

Review Section 10.2.6   Southeast Asia's Political Geography (pp. 543–546)

1.) What is Ratzel's 'Organic Theory', and how does it relate to the ideas of Political Geography today?

2.) Name the three types of geographic boundaries on p. 543, and give an example of each from the Southeast Asian realm.

3.) What are the four levels in the 'Genetic Boundary Classification' system, and where are each of these found within the realm?

A.)

B.)

C.)

D.)

4.) Name the five types of state (i.e. country) territorial configurations, and describe each. Give an example of each type found within this realm (if possible).

A.)

B.)

C.)

D.)

E.)

5.) Summarize the four 'Points to Ponder' in your own words, and why these are geopolitically and/or environmentally significant for the Southeast Asian realm.

Review Section 10.2.7   Regions: Vietnam (pp. 547–551)

1.) Name the largest cities in the Southeast Asian realm today in terms of population, and the respective country in which each is located.

2.) What are 'Doi-Moi' reforms, and how did they influence Vietnam economically after the end of the Vietnam War?

3.) How is Vietnam economically different today than it was in 1975 at the end of the Vietnam War? Give at least four examples from the textbook to support your answer.

4.) The French divided Vietnam into three units of land. Name and briefly describe these units, and how the concept of a Lingua Franca helped to unite these.

5.) What impact did the Indochina War have upon the United States, located on the other side of the Earth physically?

6.) After the end of the Indochina War and substantial migration out of the country, what steps did Vietnam take to become a global power today economically?

7.) Summarize the city of Saigon in terms of population, physical location, and cultural and economic impacts upon the country.

Review Section 10.2.8   Cambodia, Laos and Myanmar (pp. 551–557)

1.) What is 'Angkor Wat', where is it located and why is it significant in this country?

2.) What are the geographic advantages of Cambodia, as compared to other countries in the realm? Give specific examples from the textbook to support your answer.

3.) Describe the Mekong River in terms of physical location and importance to the countries of Mainland Southeast Asia, particularly Cambodia.

4.) What are some of the post-war problems associated with Cambodia, and what solutions are being created today in response to these issues?

5.) Compare Laos to the other mainland countries of Southeast Asia: How is this country similar, yet very different from the rest of the region?

6.) Describe the importance of Thailand to the realm culturally and economically. When did the transformation of Thailand into a world leader of manufacturing begin?

7.) How has tourism helped to promote economic prosperity in the realm? Give examples of specific tourism industries, and how the geographic location of Thailand has assisted the development of these.

8.) Compare Myanmar to the rest of the Southeast Asian countries today politically. What are the major differences, and how did it get to be this way?

9.) Why does Myanmar rank among the world's poorest, least-developed and most corrupt countries in the world today?

Review Section 10.2.9   Malaysia and Singapore (pp. 559–564)

1.) Describe Malaysia today in terms of unique physical location and ethnic composition.

2.) What economic improvements have been seen in Malaysia in recent years? Use the term 'Multimedia Supercorridor' in your explanation.

3.) What effect did the creation of the country of Malaysia have upon complicating the ethnic mosaic of the country today?

4.) Where is Singapore relative to Malaysia, and when did it gain independence?

5.) Compare Singapore to Malaysia today in terms of both land area and economic prosperity. Use the term 'entrepot' in your discussion.

6.) To revive its economic growth, Singapore relied upon three growth regions. Name and briefly describe each of these, and how they relate to the concept of a 'Growth Triangle'.

Review Section 10.2.10   Indonesia's Archipelago and East Timor (pp. 564–572)

1.) Name and describe the four main islands of the Indonesian Archepelago in terms of physical features, cultural diversity and economic importance.

A.)

B.)

C.)

D.)

2.) Summarize the city of Java in terms of location, population density, cultural diversity and economic importance.

3.) What is Indonesia's National Motto, and how does it relate to the tremendous cultural and language diversity found within the country?

4.) What is the concept of 'Transmigration'? Was it successful in Indonesia?

5.) Summarize the country of East Timor in terms of spatial location, population density, cultural diversity and current economic trends. When did East Timor gain independence, and from who?

Review Section 10.2.11   The Philippines (pp. 572–577)

1.) Name the three inhabited island groups of The Philippines, and describe where they are located:

A.)

B.)

C.)

2.) What religions are found in this country, and how are these creating conflicts both historically and in the present?

3.) Describe the Filipino culture in terms of origin and impact upon the country today.

4.) Where are the populations of The Philippines primarily located? What are some of the main economic activities of this country?

SECTION 10.3:   Blank Maps of Southeast Asia

Blank maps are provided for you to utilize with the mapping questions from Section 10.1 of the Study Guide. You are being provided with one blank map for each mapping exercise given in Section 10.1. It is a good idea to make additional copies of these blank maps to use for extra practice and review.

SECTION 10.4:   Student Companion Website

Additional study tools are available on the Student Companion Website at www.wiley.com/college/deblij. Features include:

- *Flashcards* offer an excellent way to practice key concepts, ideas, and terms from the text. You can review and quiz yourself on the concepts, ideas, and terms discussed in each chapter.

- *Map Quizzes* help you to master the place names for the various regions studies. Three game-formatted map activities are provided for each chapter.

- *Chapter Review Quizzes* provide immediate feedback to true/false, multiple choice, and short answer questions.

- *Audio Pronunciation* is provided for over 2000 key words and place names from the text.

- *Annotated Web Links*

- *Area and Demographic Data*

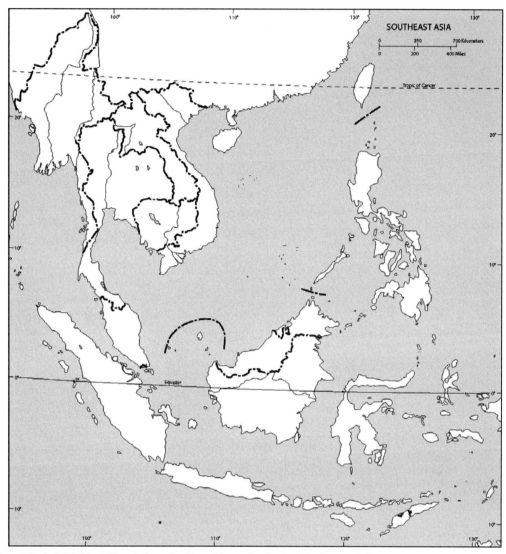

SOUTHEAST ASIA

Tropic of Cancer

Equator

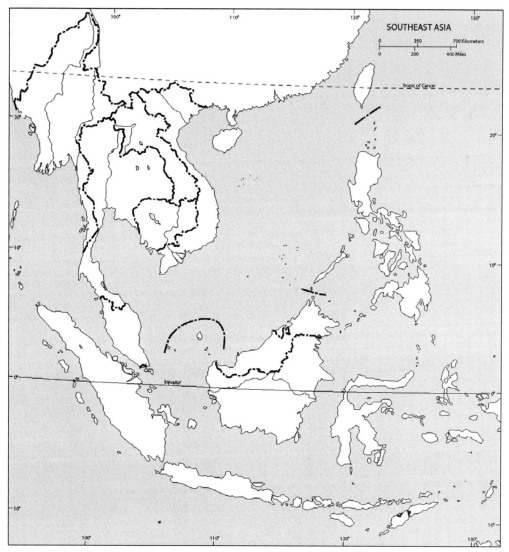

SOUTHEAST ASIA

Tropic of Cancer

Equator

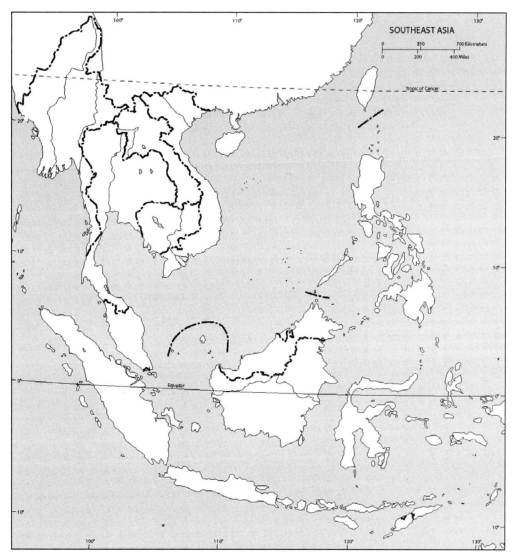

SOUTHEAST ASIA

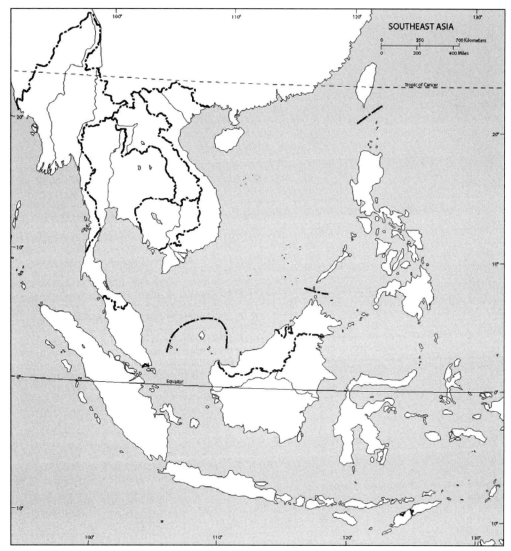

SOUTHEAST ASIA

0   350   700 Kilometers
0   200   400 Miles

Tropic of Cancer

Equator

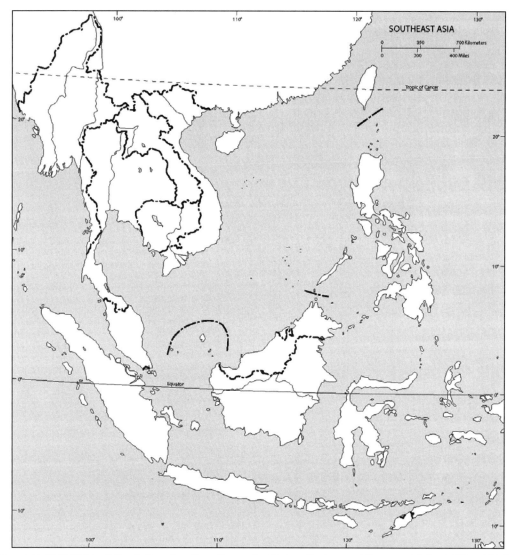

SOUTHEAST ASIA

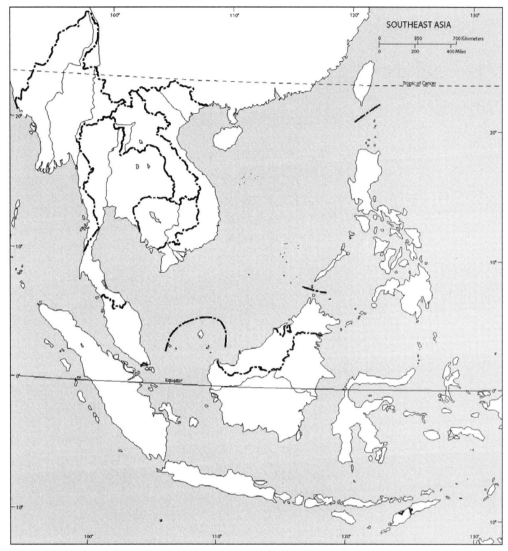

SOUTHEAST ASIA

Tropic of Cancer

Equator

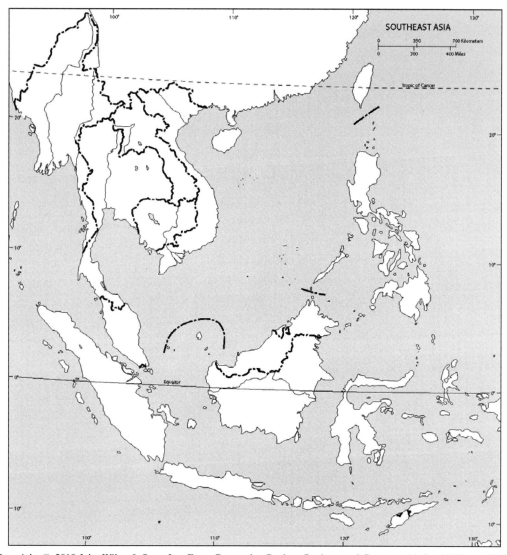

SOUTHEAST ASIA

Tropic of Cancer

Equator

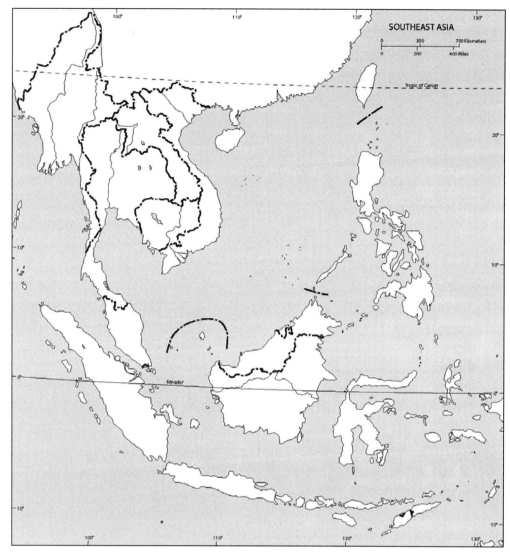

SOUTHEAST ASIA

CHAPTER #11   AUSTRALIA

SECTION 11.1:   Map Creation

In this section you will create your own detailed maps of the different major physical, cultural, geopolitical and environmental attributes of the realm. Each of these maps is critically important for understanding the Major Geographic Qualities of Australia, including their spatial distribution and functionality. By creating these on your own using the blank maps provided at the end of this chapter, you will actively learn where these different features are located and how their spatial distribution affects the ways of life in each of these respective areas. Blank Maps for Australia are provided for you in Section 11.3. Label each of your maps with the appropriate name and title. For instance, the first map you will create will be called 'Map 11–1: States of Australia'.

Map 11–1:   States of Australia

On the map, label the following seven states of Australia. To assist with creation of this map, refer to Figure 11–5, p. 586.

- Western Australia   - Northern Territory   - Queensland   - Tasmania
- New South Wales   - Australian Capital Territory   - South Australia

Map 11–2:   Major Cities of Australia

On this map, label the following major cities of the Australian Realm. You may need to use abbreviations and/or arrows if you do not have enough space to write in the full name on the map. To assist with creation of this map, refer to the overall map of Australia found on p. 578. Write in the city name using a different colored pencil or pen for each of the regions these belong.

- Perth   - Sydney   - Melbourne   - Brisbane
- Darwin   - Alice Springs   - Adelaide   - Canberra
- Hobart   - Townsville   - Weipa

Map 11–3:   Physical Features of Australia

Australia has a fairly diverse array of physical features that have influenced settlement, agricultural use, and mineral extraction throughout the realm. In this map, you will show the spatial diversity of the physical features found in Australia. To assist with creation of this map, refer to both the overall map of Australia found on p. 578, and also Figure 11–3, p. 582.

Mountain Ranges: Using a blue pen/pencil, draw in the locations of these mountain ranges on your map and label them accordingly. Use a Λ symbol to denote the locations of these features.

- Darling Range   - Great Dividing Range   - Flinders Range
- MacDonnell Ranges   - King Leopold Ranges   - Musgrave Ranges

General Physical Regions: These regions all have a physical characteristic in common, whether being basins or plateaus. Using Figure 11-4 (p. 52) as guide, circle and label the following physical regions on your map using a green pen/pencil:

- Yilgarn Plateau	- Atherton Plateau	- Kimberley Plateau
- Great Sandy Desert	- Great Victoria Desert	- Simpson Desert
- Nullarbor Basin	- Great Artesian Basin	- Murray-Darling River Basin
- Cape York Peninsula	- Eyre Peninsula	- Great Barrier Reef (offshore)

Map 11–4: Australian Bodies of Water

Australia is a realm that has been influenced by its water resources in terms of spatial history and economic development, though very few (if any) major rivers exist on the continent. On this map, you will identify the major bodies of water found inside this realm. These are divided into groups and it is recommended to use a different colored pencil or pen for each of the names per water body category. Use p. 578 as a guide.

- Darling River	- Lachlan River	- Murray River
- Pacific Ocean	- Indian Ocean	- Southern Ocean
- Coral Sea	- Great Australian Bight	- Gulf of Carpentaria
- Timor Sea	- Bass Strait	- Arafura Sea

Map 11–5: Major Climate Regions of Australia

Australia is a realm that contains three main climate types: tropical, dry, and mild mid-latitude. The major Köppen Climate types are given below for each of the climates found in this realm. Using a different colored pen/pencil for each category, outline and then color in these climate types with the World Climate Map (Figure G – 7, pp. 16–17) as a guide.

Tropical Climates:
- Af
- Aw

Dry Climates:
- BWh
- BSh
- BSk

Mild Mid-Latitude Climates:
- Cfa
- Cfb
- Csa
- Csb

High Climates:
- H

Map 11–6: Australian Population Densities

The human populations across the Australian Realm are some of the most sparse in the world, due to the continent being dominated by the deserts. Using the Australian Population Distribution map on Figure 11–2 (p. 581), we see that four general population density categories exist in the Australian Realm. Using a different colored pen/pencil for each category, outline and color the countries/sections of countries that fit into each of these designations. Once completed, you will see a spatial pattern of population distribution emerge across the realm.

Dense: This population category is only found on the eastern coast of Australia, where the primary concentration of cities is located.

Sparse: Found along the southern coastline from Sydney to Adelaide, and also the eastern half of the northern coastline.

Very Sparse: This category accounts for the rest of Australia, with most of the continent being a nearly uninhabitable desert. Small populations exist near water supplies or mining locations.

Map 11–7: Economic Activity in Australia

Using Figure 11–6 (p. 591) as a guide, circle and label the following areas on your map, displaying where these major agricultural activities are found today. Use a different color pen/pencil (if possible) for each area, and consult the textbook for further information on the situation in these locations.

- Sheep Grazing   - Cattle Grazing   - Dairy Farming
- Grain Farming   - Sugarcane Farming   - Mediterranean Agriculture
- Fishing Grounds   - Hunting/Gathering   - No Agriculture

Map 11–8: Mineral Resources in Australia

Using Figure 11–6 (p. 591) as a guide, circle and label the following main areas in which you will find each respective mining activity. Use a different color pen/pencil (if possible) for each area, and consult the textbook for further information on the situation in these locations.

- Crude Oil   - Natural Gas   - Iron Ore
- Coal Mines   - Gold Mines   - Silver Mines
- Copper Mines   - Tin Mines   - Nickel Mines

Map 11–9: Environmental Issues in Australia

The Australian Realm is one of the most environmentally-degraded places in the world today. This realm is one of the most advanced realms in the world today in terms of technology and emissions standards for air and water quality. However, moderate to severe environmental issues persist in certain locations.

These environmental issues are divided into categories below. Using a different colored pen/pencil for each, circle or highlight the areas and label them using their respective environmental situation.

Sealevel Rise:	The Great Barrier Reef, located offshore of Northeast Australia, is highly susceptible to sealevel rise. Rising waters and increased water temperatures are causing bleaching of the reef, destroying ecosystems great and small.
Coastal Pollution:	The eastern shoreline of Australia is most affected by coastal pollution, due to freight tankers coming in/out of the major ports and also from agricultural (fertilizer and pesticide) runoff into the ocean.
Groundwater Depletion:	Australia is draining their groundwater supply much faster than it can be naturally replenished, especially in the eastern regions of the continent. One potential solution is desalination, and the cities of Adelaide and Sydney are currently building such facilities to ease water shortages.
Wildfires:	The grassland regions just west of the Great Dividing Range are experiencing significant drought conditions and have been the site of some of the worst wildfire problems in the 21st century as a result. The northeastern coastline is also dealing with this problem as well.

SECTION 11.2:   Review Questions

In this section you will review the main concepts and terminology from the Australian Realm by answering the questions provided below. Write in your answers in the space provided, and refer to the respective sections/pages of the chapter for the correct information.

Review Section 11.2.1:   Australia's Major Geographic Qualities (p. 580)

Summarize the seven Major Geographic Qualities of Australia in your own words:

1.)

2.)

3.)

4.)

5.)

6.)

7.)

8.)

9.)

10.)

11.)

Review Section 11.2.2:   Land and Environment (pp. 580–584)

1.) Compare and contrast Australia with New Zealand in terms of physical regions. How are they similar, and in what major ways are they very different?

2.) How are the climates of Australia and New Zealand different from one another? What factors account for some of these differences?

3.) Where is the Southern Ocean located? What are its characteristics, and why do you not find this ocean shown on a typical map of the world?

4.) Why does Australia contain such tremendous diversity of plants and animals, unlike any other continent in the world? Give examples of these different kinds of species.

5.) Who are the 'Aboriginal' people, where are they located in the realm and when is it estimated that they arrived here?

Review Section 11.2.3:   Australia (pp. 584–596)

1.) What is the current population of Australia? What European country settled this realm, and for how long has Australia been independent?

2.) Describe Australia today in terms of development, and the effect of distance on the economy of the realm.

3.) What is the role of immigration in the population expansion of Australia today? Where do most of these immigrants come from?

4.) Where are the core and periphery areas located in Australia? How are they different physically, economically and in terms of settlement?

5.) Name the States and Federal Territories that comprise the Commonwealth of Australia today:

6.) Why would Australia be considered an 'urban culture' today? Support your answer with specific examples from the textbook.

7.) What are 'Import – Substitution Industries', and how do these relate to the economic development of Australia in the last century?

8.) List some of the major minerals found in the realm today, and the primary locations of these minerals.

9.) What is the Aboriginal Land Issue, and where in Australia is it found? What impact could this have upon the realm in terms of political boundaries?

10.) Give examples of environmental degradation in the realm, locations of these problems and their respective causes. Can you think of any solutions to these issues?

11.) Describe Australia's relationship with Indonesia and East Timor in Southeast Asia, and compare that to their relationship with Papua New Guinea in the Pacific Realm.

Review Section 11.2.4:   New Zealand (pp. 596–599)

1.) Describe New Zealand in terms of distance from Australia and physical setting. What is the current population of New Zealand today?

2.) What kinds of agricultural activities will you find in New Zealand? Give at least four examples.

3.) What factors have caused New Zealand to have peripheral development over time?

4.) Who are the Maori, and where are they located in New Zealand? Why are they creating geopolitical issues for the country?

SECTION 11.3:   Blank Maps of Australia

Blank maps are provided for you to utilize with the mapping questions from Section 11.1 of the Study Guide. You are being provided with one blank map for each mapping exercise given in Section 11.1. It is a good idea to make additional copies of these blank maps to use for extra practice and review.

SECTION 11.4:   Student Companion Website

Additional study tools are available on the Student Companion Website at www.wiley.com/college/deblij. Features include:

- *Flashcards* offer an excellent way to practice key concepts, ideas, and terms from the text. You can review and quiz yourself on the concepts, ideas, and terms discussed in each chapter.

- *Map Quizzes* help you to master the place names for the various regions studies. Three game-formatted map activities are provided for each chapter.

- *Chapter Review Quizzes* provide immediate feedback to true/false, multiple choice, and short answer questions.

- *Audio Pronunciation* is provided for over 2000 key words and place names from the text.

- *Annotated Web Links*

- *Area and Demographic Data*

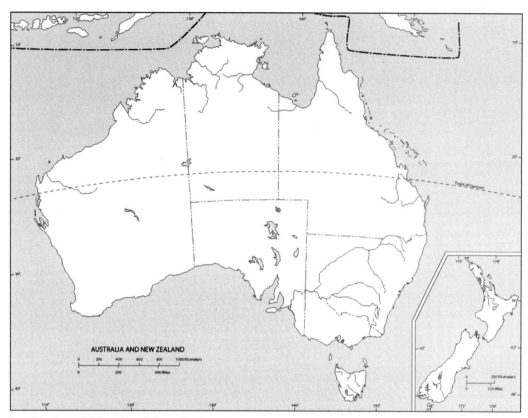

AUSTRALIA AND NEW ZEALAND

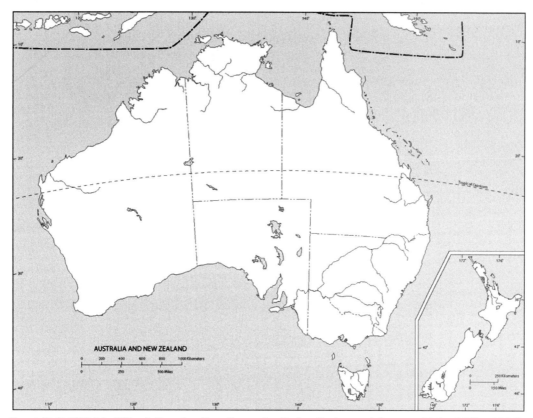

AUSTRALIA AND NEW ZEALAND

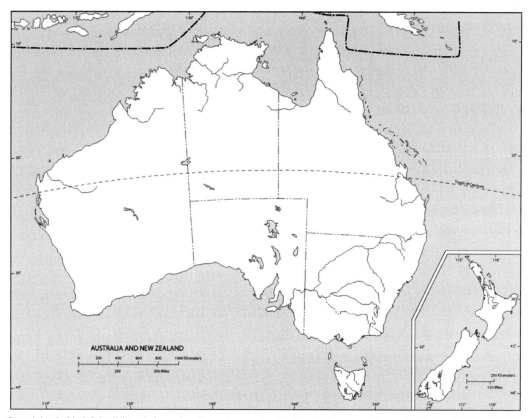

AUSTRALIA AND NEW ZEALAND

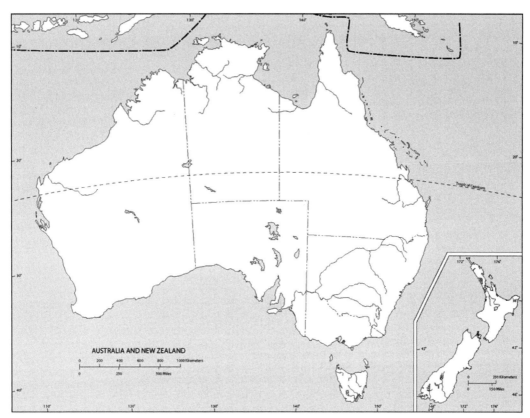

AUSTRALIA AND NEW ZEALAND

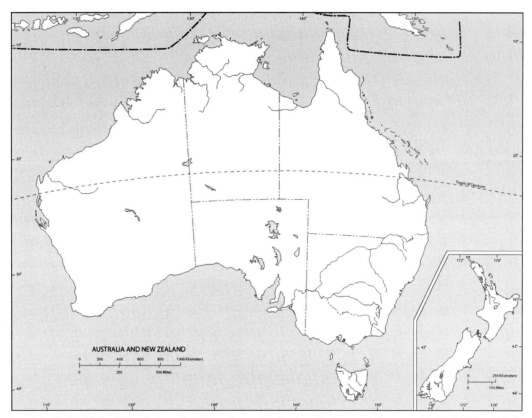

AUSTRALIA AND NEW ZEALAND

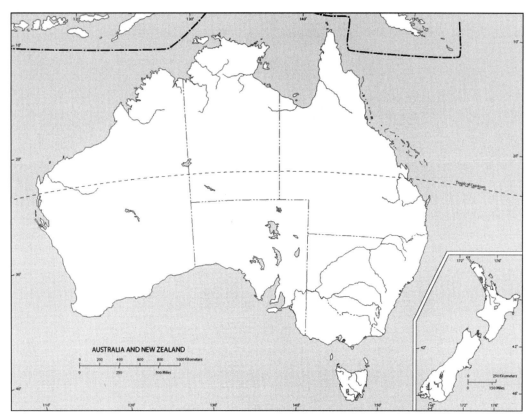

AUSTRALIA AND NEW ZEALAND

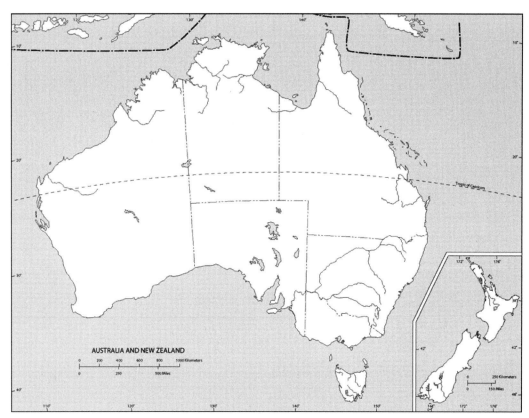

AUSTRALIA AND NEW ZEALAND

CHAPTER #12   THE PACIFIC REALM

SECTION 12.1:   Map Creation

In this section you will create your own detailed maps of the different major physical, cultural, geopolitical and environmental attributes of the realm. Each of these maps is critically important for understanding the Major Geographic Qualities of The Pacific, including their spatial distribution and functionality. By creating these on your own using the blank maps provided at the end of this chapter, you will actively learn where these different features are located and how their spatial distribution affects the ways of life in each of these respective areas. Blank Maps for The Pacific are provided for you in Section 12.3. Label each of your maps with the appropriate name and title. For instance, the first map you will create will be called 'Map 12–1: Countries of the Pacific Realm'.

Map 12–1:   Countries of the Pacific Realm

On the map, label the following countries of the Pacific Realm. These have been divided into three regions, and you may need to use abbreviations and/or arrows if you do not have enough space to write in the full name on the map. To assist with creation of this map, refer to Figure 12–2, pp. 604–605.

- New Caledonia	- Vanuatu	- Solomon Islands	- Papua New Guinea
- Nauru	- Palau	- Guam	- Mariana Islands
- Marshall Islands	- Kiribati	- Tuvalu	- Fiji
- Midway Islands	- Samoa	- Tahiti	- French Polynesia

Map 12–2:   Major Climate Regions of The Pacific

The Pacific is a realm that contains only two climate types: Tropical and High Climates. Use the World Climate Map (Figure G – 8, Introductory Section) as your guide. The major Köppen Climate types are given below for each of the climates found in this realm. Using a different colored pen/pencil for each category, outline and then color in these climate types:

Tropical Climates:
  - Af (99% of the Pacific)
  - Aw (Southern Pacific)

High Climates:
  - H (Papua New Guinea)

Map 12–3:   Geopolitical Issues in The Pacific Realm

In this map you will show the locations of geopolitical conflicts within The Pacific Realm. After completing this exercise, consult the textbook to understand why these situations are confronting various regions of the realm today.

Island of Fiji: Ethnic tensions, and a military coup in 1999

Papua New Guinea: Several Geopolitical Issues here:
  • Civil war and unrest between ethnic groups
  • Problems with disease (malaria and HIV/AIDS)
  • An island of Papua New Guinea wants independence; if granted, they would control most of the copper mines in the country

Marshall Islands:	Nuclear testing site for the French
Tahiti:	Overuse of the island for tourism; low-paying jobs for most of the local people as Europeans/Americans are flown in to handle customer service positions.

Map 12–4: *Environmental Issues in The Pacific Realm*

The Pacific Realm contains some environmental issues of concern today, even with it occupying nearly an entire hemisphere with very little population density in comparison to other realms. These environmental issues are divided into categories below. Using a different colored pen/pencil for each, circle or highlight the areas and label them using their respective environmental situation.

Sealevel Rise:	The Pacific Islands with low-lying elevations such as the Solomon Islands are the highly susceptible to sealevel rise, as some of these islands are only 4 feet above sea level. These are found throughout the Northwest and Southwestern regions of the realm.
Nuclear Testing:	Most of the Central Pacific has been a site for nuclear testing by the developed nations of the world, and some areas remain so today. Radiation from nuclear fallout can severely disrupt wildlife populations and ecosystems, both on land and in the ocean waters.
Deforestation:	This is occurring at very high levels in island countries where tourism is the primary economic activity, as land is being cleared for hotel rooms and permanent residences. Additionally, Papua New Guinea is seeing high levels of population increases due to a high birth rate, and so deforestation is occurring there in favor of land for agriculture and homes for these people.
Geologic Hazards:	All of the Pacific Realm is under the threat of a tsunami, should one occur (in the likely epicenter region of the South Pacific or Southeast Asia). Strong earthquake zones and active volcanoes line the region to the west and south.

Map 12–5: *Regions of The Pacific Realm*

Using Figure 6–11 (p. 303) in the textbook, color the following regions of the Sub-Saharan Africa Realm and use a different colored pen/pencil for each region and the country list below as a guide. After completing this exercise, consult the textbook to understand why each of these is considered a separate region of the realm.

Micronesia:	Meaning 'small islands', these are found mainly in the northwest Pacific.
Melanesia:	Meaning 'dark islands', these are found along the equator to the South Pacific, excluding the Eastern Pacific areas.
Polynesia:	Meaning 'many islands', this region is located in the Eastern Pacific.

SECTION 12.2:   Review Questions

In this section you will review the main concepts and terminology from The Pacific Realm by answering the questions provided below. Write in your answers in the space provided, and refer to the respective sections/pages of the chapter for the correct information.

Review Section 12.2.1   The Pacific Realm's Major Geographic Qualities (p. 602)

Summarize the nine Major Geographic Qualities of The Pacific in your own words:

1.)

2.)

3.)

4.)

5.)

6.)

7.)

8.)

9.)

10.)

11.)

Review Section 12.2.2   Defining the Realm (pp. 602–603)

1.) Where is the Pacific Realm located, and how large of an area of Earth does it occupy? What is its present population?

2.) Compare the Arctic to the Antarctic regions, describing why these two areas are vastly different.

3.) What is the effect of climate change in this realm? Give specific examples to support your answer.

4.) Why are Australia and New Zealand not considered to be a part of this realm?

Review Section 12.2.3   Colonization and Independence (p. 603)

1.) What European countries colonized this realm? What regions of the realm did they occupy?

Review Section 12.2.4   The Pacific Realm's Marine Geography (pp. 603–607)

1.) What is the difference between the 'territorial sea' and the 'high seas'? How do these definitions apply to the 'Scramble for the Seas' starting in 1945?

2.) What were the spatial results of the UNCLOS Intervention? Use the terms 'territorial sea' and 'exclusive economic zone' in your explanation.

3.) Describe the 'Maritime Boundary' problems associated with the UNCLOS Intervention, and the role of 'median lines' as a form of resolution to these issues.

4.) What are some of the EEZ implications for countries with a stake in land in The Pacific Realm?

Review Section 12.2.5   Regions of the Realm (pp. 608–614)

1.) List and describe the three regions of The Pacific Realm, contrasting these with one another in terms of countries included within the region, population, location and cultural differences:

A.)

B.)

C.)

2.) Besides Honolulu (Hawaii, USA), name the three largest cities in the realm and their respective populations. What does this imply about the population density throughout the realm?

3.) From the discussion on p. 610 ('Who Should Own the Oceans?'), what are the pros and cons of the argument being presented? What is your opinion on the situation?

Review Section 12.2.7   The Antarctic Partition (pp. 614–616)

1.) How has Antarctica been divided between countries of the world, and what nations have staked claims in this continent? Why?

Review Section 12.2.8   Geopolitics in the Arctic (pp. 616–619)

1.) What are some of the current geopolitical issues that face the Arctic regions? What solutions can you think of for these situations?

SECTION 12.3:   Blank Maps of The Pacific Realm

Blank maps are provided for you to utilize with the mapping questions from Section 12.1 of the Study Guide. You are being provided with one blank map for each mapping exercise given in Section 12.1. It is a good idea to make additional copies of these blank maps to use for extra practice and review.

SECTION 1.4:   Student Companion Website

Additional study tools are available on the Student Companion Website at www.wiley.com/college/ deblij. Features include:

- *Flashcards* offer an excellent way to practice key concepts, ideas, and terms from the text. You can review and quiz yourself on the concepts, ideas, and terms discussed in each chapter.

- *Map Quizzes* help you to master the place names for the various regions studies. Three game-formatted map activities are provided for each chapter.

- *Chapter Review Quizzes* provide immediate feedback to true/false, multiple choice, and short answer questions.

- *Audio Pronunciation* is provided for over 2000 key words and place names from the text.

- *Annotated Web Links*

- *Area and Demographic Data*

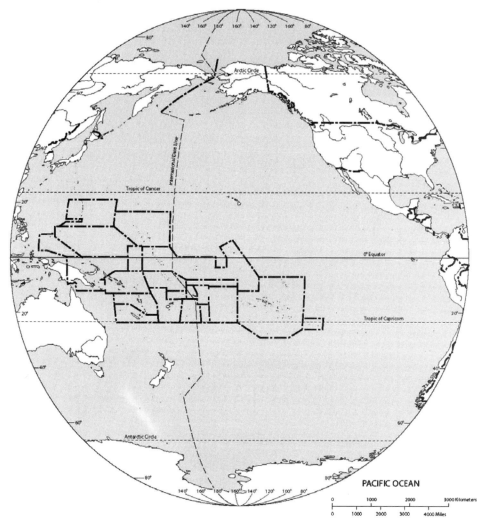

PACIFIC OCEAN

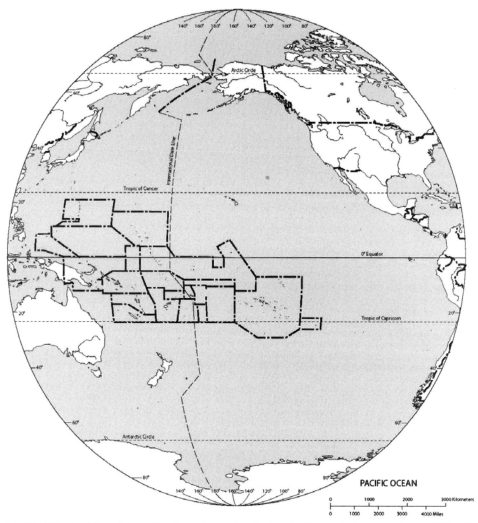

PACIFIC OCEAN

0 1000 2000 3000 Kilometers

0 1000 2000 3000 4000 Miles

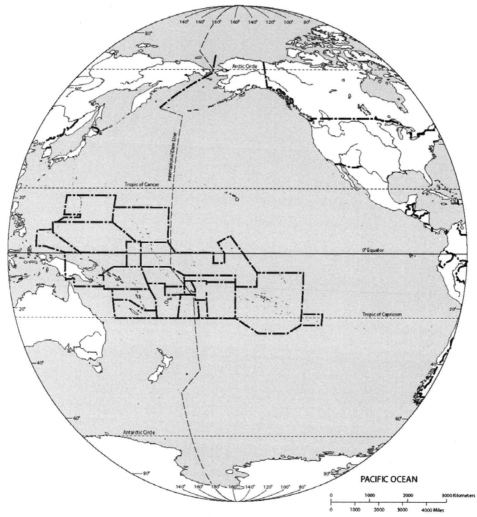

PACIFIC OCEAN

0   1000   2000   3000 Kilometers
0   1000   2000   3000   4000 Miles

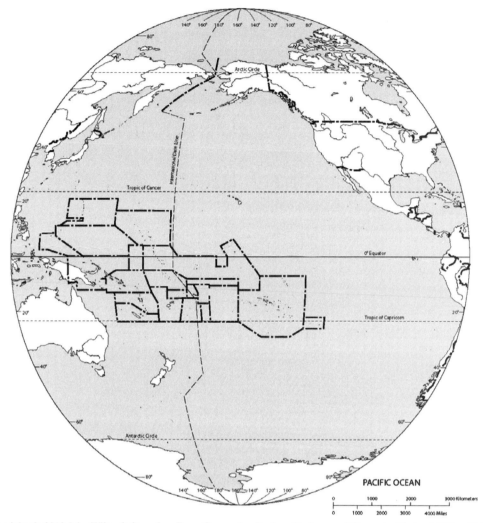

PACIFIC OCEAN

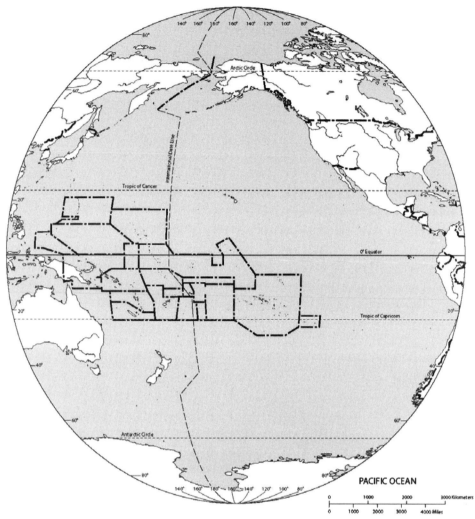

PACIFIC OCEAN